Holt Mathematics

Course 3
Lesson Plans

HOLT, RINEHART AND WINSTON
A Harcourt Education Company

Orlando • Austin • New York • San Diego • Toronto • London

Copyright © by Holt, Rinehart and Winston

All rights reserved. No part of this publication may be reproduced or transmitted in any form or by any means, electronic or mechanical, including photocopy, recording, or any information storage and retrieval system, without permission in writing from the publisher.

Teachers using HOLT MATHEMATICS may photocopy complete pages in sufficient quantities for classroom use only and not for resale.

HOLT and the **"Owl Design"** are trademarks licensed to Holt, Rinehart and Winston, registered in the United States of America and/or other jurisdictions.

Printed in the United States of America

If you have received these materials as examination copies free of charge, Holt, Rinehart and Winston retains title to the materials and they may not be resold. Resale of examination copies is strictly prohibited.

Possession of this publication in print format does not entitle users to convert this publication, or any portion of it, into electronic format.

ISBN 0-03-078473-5
6 7 8 9 10   170   12 11 10 09

Contents

Chapter 1
Lesson 1-1 1
Lesson 1-2 2
Lesson 1-3 3
Lesson 1-4 4
Lesson 1-5 5
Lesson 1-6 6
Lesson 1-7 7
Lesson 1-8 8
Lesson 1-9 9

Chapter 2
Lesson 2-1 10
Lesson 2-2 11
Lesson 2-3 12
Lesson 2-4 13
Lesson 2-5 14
Lesson 2-6 15
Lesson 2-7 16
Lesson 2-8 17

Chapter 3
Lesson 3-1 18
Lesson 3-2 19
Lesson 3-3 20
Lesson 3-4 21
Lesson 3-5 22
Lesson 3-6 23

Chapter 4
Lesson 4-1 24
Lesson 4-2 25
Lesson 4-3 26
Lesson 4-4 27
Lesson 4-5 28
Lesson 4-6 29
Lesson 4-7 30
Lesson 4-8 31

Chapter 5
Lesson 5-1 32
Lesson 5-2 33
Lesson 5-3 34
Lesson 5-4 35
Lesson 5-5 36
Lesson 5-6 37
Lesson 5-7 38
Lesson 5-8 39

Chapter 6
Lesson 6-1 40
Lesson 6-2 41
Lesson 6-3 42
Lesson 6-4 43
Lesson 6-5 44

Lesson 6-6 45
Lesson 6-7 46

Chapter 7
Lesson 7-1 47
Lesson 7-2 48
Lesson 7-3 49
Lesson 7-4 50
Lesson 7-5 51
Lesson 7-6 52
Lesson 7-7 53
Lesson 7-8 54
Lesson 7-9 55

Chapter 8
Lesson 8-1 56
Lesson 8-2 57
Lesson 8-3 58
Lesson 8-4 59
Lesson 8-5 60
Lesson 8-6 61
Lesson 8-7 62
Lesson 8-8 63
Lesson 8-9 64
Lesson 8-10 65

Chapter 9
Lesson 9-1 66
Lesson 9-2 67
Lesson 9-3 68
Lesson 9-4 69
Lesson 9-5 70
Lesson 9-6 71
Lesson 9-7 72
Lesson 9-8 73

Chapter 10
Lesson 10-1 74
Lesson 10-2 75
Lesson 10-3 76
Lesson 10-4 77
Lesson 10-5 78
Lesson 10-6 79
Lesson 10-7 80
Lesson 10-8 81
Lesson 10-9 82

Chapter 11
Lesson 11-1 83
Lesson 11-2 84
Lesson 11-3 85
Lesson 11-4 86
Lesson 11-5 87
Lesson 11-6 88

Contents

Chapter 12
Lesson 12-1 89
Lesson 12-2 90
Lesson 12-3 91
Lesson 12-4 92
Lesson 12-5 93
Lesson 12-6 94
Lesson 12-7 95

Chapter 13
Lesson 13-1 96
Lesson 13-2 97
Lesson 13-3 98
Lesson 13-4 99
Lesson 13-5 100
Lesson 13-6 101
Lesson 13-7 102

Chapter 14
Lesson 14-1 103
Lesson 14-2 104
Lesson 14-3 105
Lesson 14-4 106
Lesson 14-5 107
Lesson 14-6 108

Teacher's Name _____ Class _____ Date _____

Lesson Plan 1-1
Variables and Expressions pp. 6–9 Day _____

Objective Students evaluate algebraic expressions.

> **NCTM Standards:** Understand meanings of operations and how they relate to one another; Represent and analyze mathematical situations and structures using algebraic symbols.

Pacing
☐ 45-minute Classes: 1 day   ☐ 90-minute Classes: 1/2 day   ☐ Other_____

WARM UP
☐ Warm Up TE p. 6 and Daily Transparency 1-1
☐ Problem of the Day TE p. 6 and Daily Transparency 1-1
☐ Countdown to Testing Transparency Week 1

TEACH
☐ Lesson Presentation CD-ROM 1-1
☐ Alternate Opener, Explorations Transparency 1-1, TE p. 6, and Exploration 1-1
☐ Reaching All Learners TE p. 7
☐ *Know-It Notebook* 1-1

PRACTICE AND APPLY
☐ Example 1: Average: 1–3, 10–12, 19–23 odd, 53–67 Advanced: 10–12, 21–29 odd, 52–67
☐ Example 2: Average: 1–5, 10–14, 19–45 odd, 53–67 Advanced: 10–14, 48–50, 52–67
☐ Example 3: Average: 1–18, 43–47 odd, 53–67 Advanced: 11–17 odd, 19–67

REACHING ALL LEARNERS – Differentiated Instruction for students with

Developing Knowledge	On-level Knowledge	Advanced Knowledge	English Language Development
☐ Cooperative Learning TE p. 7	☐ Cooperative Learning TE p. 7	☐ Cooperative Learning TE p. 7	☐ Cooperative Learning TE p. 7
☐ Practice A 1-1 CRB	☐ Practice B 1-1 CRB	☐ Practice C 1-1 CRB	☐ Practice A, B, or C 1-1 CRB
☐ Reteach 1-1 CRB	☐ Puzzles, Twisters & Teasers 1-1 CRB	☐ Challenge 1-1 CRB	☐ *Success for ELL* 1-1
☐ Homework Help Online Keyword: MT7 1-1	☐ Homework Help Online Keyword: MT7 1-1	☐ Homework Help Online Keyword: MT7 1-1	☐ Homework Help Online Keyword: MT7 1-1
☐ *Lesson Tutorial Video* 1-1	☐ *Lesson Tutorial Video* 1-1	☐ *Lesson Tutorial Video* 1-1	☐ *Lesson Tutorial Video* 1-1
☐ Reading Strategies 1-1 CRB	☐ Problem Solving 1-1 CRB	☐ Problem Solving 1-1 CRB	☐ Reading Strategies 1-1 CRB
☐ *Questioning Strategies* p. 1	☐ Reading Math TE p. 7	☐ Reading Math TE p. 7	☐ Lesson Vocabulary SE p. 6
☐ *IDEA Works!* 1-1			☐ *Multilingual Glossary*

ASSESSMENT
☐ Lesson Quiz, TE p. 9 and DT 1-1   ☐ State-Specific Test Prep Online Keyword: MT7 TestPrep

Holt Mathematics

Teacher's Name _____ Class _____ Date _____

Lesson Plan 1-2
Algebraic Expressions pp. 10–13   Day _____

Objective Students write algebraic equations.

> **NCTM Standards:** Understand meanings of operations and how they relate to one another; Represent and analyze mathematical situations and structures using algebraic symbols;

Pacing
- ☐ 45-minute Classes: 1 day   ☐ 90-minute Classes: 1/2 day   ☐ Other_____

WARM UP
- ☐ Warm Up TE p. 10 and Daily Transparency 1-2
- ☐ Problem of the Day TE p. 10 and Daily Transparency 1-2
- ☐ Countdown to Testing Transparency Week 1

TEACH
- ☐ Lesson Presentation CD-ROM 1-2
- ☐ Alternate Opener, Explorations Transparency 1-2, TE p. 10, and Exploration 1-2
- ☐ Reaching All Learners TE p. 11
- ☐ Teaching Transparency 1-2
- ☐ *Know-It Notebook* 1-2

PRACTICE AND APPLY
- ☐ Example 1: Average: 1–4, 11–14, 23–31 odd, 38–48 Advanced: 11–15, 22–31, 37–48
- ☐ Example 2: Average: 1–8, 11–19, 28–35, 38–48 Advanced: 11–19, 28–35, 38–48
- ☐ Example 3: Average: 1–9, 11–20, 38–48 Advanced: 11–20, 36–48
- ☐ Example 4: Average: 1–21, 38–48 Advanced: 11–21, 36–48

REACHING ALL LEARNERS – Differentiated Instruction for students with

Developing Knowledge	On-level Knowledge	Advanced Knowledge	English Language Development
☐ Kinesthetic Experience TE p. 11	☐ Kinesthetic Experience TE p. 11	☐ Kinesthetic Experience TE p. 11	☐ Kinesthetic Experience TE p. 11
☐ Practice A 1-2 CRB	☐ Practice B 1-2 CRB	☐ Practice C 1-2 CRB	☐ Practice A, B, or C 1-2 CRB
☐ Reteach 1-2 CRB	☐ Puzzles, Twisters & Teasers 1-2 CRB	☐ Challenge 1-2 CRB	☐ *Success for ELL* 1-2
☐ Homework Help Online Keyword: MT7 1-2	☐ Homework Help Online Keyword: MT7 1-2	☐ Homework Help Online Keyword: MT7 1-2	☐ Homework Help Online Keyword: MT7 1-2
☐ *Lesson Tutorial Video* 1-2	☐ *Lesson Tutorial Video* 1-2	☐ *Lesson Tutorial Video* 1-2	☐ *Lesson Tutorial Video* 1-2
☐ Reading Strategies 1-2 CRB	☐ Problem Solving 1-2 CRB	☐ Problem Solving 1-2 CRB	☐ Reading Strategies 1-2 CRB
☐ *Questioning Strategies* pp. 2–3			
☐ *IDEA Works!* 1-2			☐ *Multilingual Glossary*

ASSESSMENT
- ☐ Lesson Quiz, TE p. 13 and DT 1-2   ☐ State-Specific Test Prep Online Keyword: MT7 TestPrep

Teacher's Name _____ Class _____ Date _____

Lesson Plan 1-3
Integers and Absolute Value pp. 14–17   Day _____

Objective Students compare and order integers, and evaluate expressions containing absolute values.

> **NCTM Standards:** Understand numbers, ways of representing numbers, relationships among numbers, and number systems; Compute fluently and make reasonable estimates; Create and use representations to organize, record, and communicate mathematical ideas.

Pacing
☐ 45-minute Classes: 1 day   ☐ 90-minute Classes: 1/2 day   ☐ Other_____

WARM UP
☐ Warm Up TE p. 14 and Daily Transparency 1-3
☐ Problem of the Day TE p. 14 and Daily Transparency 1-3
☐ Countdown to Testing Transparency Week 1

TEACH
☐ Lesson Presentation CD-ROM 1-3
☐ Alternate Opener, Explorations Transparency 1-3, TE p. 14, and Exploration 1-3
☐ Reaching All Learners TE p. 15
☐ Teaching Transparency 1-3
☐ *Know-It Notebook* 1-3

PRACTICE AND APPLY
☐ Example 1: Average: 1, 15, 29–36, 57–65 Advanced: 15, 29–36, 52, 57–65
☐ Example 2: Average: 1–5, 15–19, 29–39 Advanced: 15–19, 29–39, 52–53, 57–65
☐ Example 3: Average: 1–10, 29–39, 57–65 Advanced: 15–24, 29–39, 48–53, 57–65
☐ Example 4: Average: 1–13 odd, 15–28, 29–51 odd, 57–65 Advanced: 15–27 odd, 29–51 odd, 52–65

REACHING ALL LEARNERS – Differentiated Instruction for students with

Developing Knowledge	On-level Knowledge	Advanced Knowledge	English Language Development
☐ Graphic Organizers TE p. 15	☐ Graphic Organizers TE p. 15	☐ Graphic Organizers TE p. 15	☐ Graphic Organizers TE p. 15
☐ Practice A 1-3 CRB	☐ Practice B 1-3 CRB	☐ Practice C 1-3 CRB	☐ Practice A, B, or C 1-3 CRB
☐ Reteach 1-3 CRB	☐ Puzzles, Twisters & Teasers 1-3 CRB	☐ Challenge 1-3 CRB	☐ *Success for ELL* 1-3
☐ Homework Help Online Keyword: MT7 1-3	☐ Homework Help Online Keyword: MT7 1-3	☐ Homework Help Online Keyword: MT7 1-3	☐ Homework Help Online Keyword: MT7 1-3
☐ *Lesson Tutorial Video* 1-3	☐ *Lesson Tutorial Video* 1-3	☐ *Lesson Tutorial Video* 1-3	☐ *Lesson Tutorial Video* 1-3
☐ Reading Strategies 1-3 CRB	☐ Problem Solving 1-3 CRB	☐ Problem Solving 1-3 CRB	☐ Reading Strategies 1-3 CRB
☐ *Questioning Strategies* pp. 4–5	☐ Visual TE p. 15	☐ Visual TE p. 15	☐ Lesson Vocabulary SE p. 14
☐ *IDEA Works!* 1-3			☐ *Multilingual Glossary*

ASSESSMENT
☐ Lesson Quiz, TE p. 17 and DT 1-3   ☐ State-Specific Test Prep Online Keyword: MT7 TestPrep

Teacher's Name _____ Class _____ Date _____

Lesson Plan 1-4
Adding Integers pp. 18–21   Day _____

Objective Students add integers.

> **NCTM Standards:** Compute fluently and make reasonable estimates.

Pacing
☐ 45-minute Classes: 1 day   ☐ 90-minute Classes: 1/2 day   ☐ Other_____

WARM UP
☐ Warm Up TE p. 18 and Daily Transparency 1-4
☐ Problem of the Day TE p. 18 and Daily Transparency 1-4
☐ Countdown to Testing Transparency Week 1

TEACH
☐ Lesson Presentation CD-ROM 1-4
☐ Alternate Opener, Explorations Transparency 1-4, TE p. 18, and Exploration 1-4
☐ Reaching All Learners TE p. 19
☐ Teaching Transparency 1-4
☐ *Hands-On Lab Activities* 1-4
☐ *Know-It Notebook* 1-4

PRACTICE AND APPLY
☐ Example 1: Average: 1–4, 13–16, 29–38 odd, 49–56 Advanced: 13–16, 29–38, 48–56
☐ Example 2: Average: 1–8, 13–20, 31–38, 49–56 Advanced: 13–20, 48–56
☐ Example 3: Average: 1–11, 13–27, 39–44, 49–56 Advanced: 13–27, 46, 48–56
☐ Example 4: Average: 1–28, 29–43 odd, 49–56 Advanced: 13–45 odd, 47–56

REACHING ALL LEARNERS – Differentiated Instruction for students with

Developing Knowledge	On-level Knowledge	Advanced Knowledge	English Language Development
☐ Inclusion TE p. 19	☐ Concrete Manipulatives TE p. 19	☐ Concrete Manipulatives TE p. 19	☐ Concrete Manipulatives TE p. 19
☐ Practice A 1-4 CRB	☐ Practice B 1-4 CRB	☐ Practice C 1-4 CRB	☐ Practice A, B, or C 1-4 CRB
☐ Reteach 1-4 CRB	☐ Puzzles, Twisters & Teasers 1-4 CRB	☐ Challenge 1-4 CRB	☐ *Success for ELL* 1-4
☐ Homework Help Online Keyword: MT7 1-4	☐ Homework Help Online Keyword: MT7 1-4	☐ Homework Help Online Keyword: MT7 1-4	☐ Homework Help Online Keyword: MT7 1-4
☐ *Lesson Tutorial Video* 1-4	☐ *Lesson Tutorial Video* 1-4	☐ *Lesson Tutorial Video* 1-4	☐ *Lesson Tutorial Video* 1-4
☐ Reading Strategies 1-4 CRB	☐ Problem Solving 1-4 CRB	☐ Problem Solving 1-4 CRB	☐ Reading Strategies 1-4 CRB
☐ *Questioning Strategies* pp. 6–7			
☐ *IDEA Works!* 1-4			☐ *Multilingual Glossary*

ASSESSMENT
☐ Lesson Quiz, TE p. 21 and DT 1-4   ☐ State-Specific Test Prep Online Keyword: MT7 TestPrep

Copyright © Holt, Rinehart and Winston.
All rights reserved.

Holt Mathematics

Teacher's Name _____ Class _____ Date _____

Lesson Plan 1-5
Subtracting Integers pp. 22–25   Day _____

Objective Students subtract integers.

> **NCTM Standards:** Compute fluently and make reasonable estimates; Recognize and apply mathematics in contexts outside of mathematics.

Pacing
- [ ] 45-minute Classes: 1 day   - [ ] 90-minute Classes: 1/2 day   - [ ] Other _____

WARM UP
- [ ] Warm Up TE p. 22 and Daily Transparency 1-5
- [ ] Problem of the Day TE p. 22 and Daily Transparency 1-5
- [ ] Countdown to Testing Transparency Week 1

TEACH
- [ ] Lesson Presentation CD-ROM 1-5
- [ ] Alternate Opener, Explorations Transparency 1-5, TE p. 22, and Exploration 1-5
- [ ] Reaching All Learners TE p. 23
- [ ] Teaching Transparency 1-5
- [ ] *Know-It Notebook* 1-5

PRACTICE AND APPLY
- [ ] Example 1: Average: 1–4, 9–12, 19–24, 36–42   Advanced: 9–12, 17–24, 30–31, 36–42
- [ ] Example 2: Average: 1–7, 9–15, 19–28, 36–42   Advanced: 9–15, 17–28, 29–34, 36–42
- [ ] Example 3: Average: 1–29 odd, 30–42   Advanced: 9–29 odd, 30–42

REACHING ALL LEARNERS – Differentiated Instruction for students with

Developing Knowledge	On-level Knowledge	Advanced Knowledge	English Language Development
☐ Inclusion p. 23	☐ Critical Thinking TE p. 23	☐ Critical Thinking TE p. 23	☐ Critical Thinking TE p. 23
☐ Practice A 1-5 CRB	☐ Practice B 1-5 CRB	☐ Practice C 1-5 CRB	☐ Practice A, B, or C 1-5 CRB
☐ Reteach 1-5 CRB	☐ Puzzles, Twisters & Teasers 1-5 CRB	☐ Challenge 1-5 CRB	☐ *Success for ELL* 1-5
☐ Homework Help Online Keyword: MT7 1-5	☐ Homework Help Online Keyword: MT7 1-5	☐ Homework Help Online Keyword: MT7 1-5	☐ Homework Help Online Keyword: MT7 1-5
☐ *Lesson Tutorial Video* 1-5	☐ *Lesson Tutorial Video* 1-5	☐ *Lesson Tutorial Video* 1-5	☐ *Lesson Tutorial Video* 1-5
☐ Reading Strategies 1-5 CRB	☐ Problem Solving 1-5 CRB	☐ Problem Solving 1-5 CRB	☐ Reading Strategies 1-5 CRB
☐ *Questioning Strategies* pp. 8–9			
☐ *IDEA Works!* 1-5			☐ *Multilingual Glossary*

ASSESSMENT
- [ ] Lesson Quiz, TE p. 25 and DT 1-5   - [ ] State-Specific Test Prep Online Keyword: MT7 TestPrep

Teacher's Name _____ Class _____ Date _____

Lesson Plan 1-6
Multiplying and Dividing Integers pp. 26–29   Day _____

Objective Students multiply and divide integers.

> **NCTM Standards:** Compute fluently and make reasonable estimates; Recognize and apply mathematics in contexts outside of mathematics.

Pacing
☐ 45-minute Classes: 1 day   ☐ 90-minute Classes: 1/2 day   ☐ Other_____

WARM UP
☐ Warm Up TE p. 26 and Daily Transparency 1-6
☐ Problem of the Day TE p. 26 and Daily Transparency 1-6
☐ Countdown to Testing Transparency Week 2

TEACH
☐ Lesson Presentation CD-ROM 1-6
☐ Alternate Opener, Explorations Transparency 1-6, TE p. 26, and Exploration 1-6
☐ Reaching All Learners TE p. 27
☐ Teaching Transparency 1-6
☐ *Technology Lab Activities* 1-6
☐ *Know-It Notebook* 1-6

PRACTICE AND APPLY
☐ Example 1: Average: 1–4, 11–14, 22–29, 41–49 Advanced: 11–14, 22–35, 41–49
☐ Example 2: Average: 1–8, 11–18, 23–35 odd, 41–49 Advanced: 11–18, 22–36, 41–49
☐ Example 3: Average: 1–20, 23–35 odd, 38–49 Advanced: 11–21, 30–49

REACHING ALL LEARNERS – Differentiated Instruction for students with

Developing Knowledge	On-level Knowledge	Advanced Knowledge	English Language Development
☐ Critical Thinking TE p. 27	☐ Critical Thinking TE p. 27	☐ Critical Thinking TE p. 27	☐ Critical Thinking TE p. 27
☐ Practice A 1-6 CRB	☐ Practice B 1-6 CRB	☐ Practice C 1-6 CRB	☐ Practice A, B, or C 1-6 CRB
☐ Reteach 1-6 CRB	☐ Puzzles, Twisters & Teasers 1-6 CRB	☐ Challenge 1-6 CRB	☐ *Success for ELL* 1-6
☐ Homework Help Online Keyword: MT7 1-6	☐ Homework Help Online Keyword: MT7 1-6	☐ Homework Help Online Keyword: MT7 1-6	☐ Homework Help Online Keyword: MT7 1-6
☐ *Lesson Tutorial Video* 1-6	☐ *Lesson Tutorial Video* 1-6	☐ *Lesson Tutorial Video* 1-6	☐ *Lesson Tutorial Video* 1-6
☐ Reading Strategies 1-6 CRB	☐ Problem Solving 1-6 CRB	☐ Problem Solving 1-6 CRB	☐ Reading Strategies 1-6 CRB
☐ *Questioning Strategies* pp. 10–11	☐ Communicating Math TE p. 27	☐ Communicating Math TE p. 27	
☐ *IDEA Works!* 1-6			☐ *Multilingual Glossary*

ASSESSMENT
☐ Lesson Quiz, TE p. 29 and DT 1-6   ☐ State-Specific Test Prep Online Keyword: MT7 TestPrep

Copyright © Holt, Rinehart and Winston.
All rights reserved.

Holt Mathematics

Teacher's Name _____ Class _____ Date _____

Lesson Plan 1-7
Solving Equations by Adding or Subtracting pp. 34–38 Day _____

Objective Students solve equations using addition and subtraction.

> **NCTM Standards:** Understand patterns, relations, and functions; Recognize reasoning and proof as fundamental aspects of mathematics; Make and investigate mathematical conjectures; Select and use various types of reasoning and methods of proof.

Pacing
☐ 45-minute Classes: 1 day ☐ 90-minute Classes: 1/2 day ☐ Other_____

WARM UP
☐ Warm Up TE p. 34 and Daily Transparency 1-7
☐ Problem of the Day TE p. 34 and Daily Transparency 1-7
☐ Countdown to Testing Transparency Week 2

TEACH
☐ Lesson Presentation CD-ROM 1-7
☐ Alternate Opener, Explorations Transparency 1-7, TE p. 34, and Exploration 1-7
☐ Reaching All Learners TE p. 35
☐ Teaching Transparency 1-7
☐ *Hands-On Lab Activities* 1-7
☐ *Know-It Notebook* 1-7

PRACTICE AND APPLY
☐ Example 1: Average: 1, 2, 10, 11, 36–41, 46–56 Advanced: 10, 11, 36–41, 46–56
☐ Example 2: Average: 1–8, 10–17, 19–41 odd, 46–56 Advanced: 10–17, 28–42, 45–56
☐ Example 3: Average: 1–17 odd, 42–56 Advanced: 15–18, 42–56

REACHING ALL LEARNERS – Differentiated Instruction for students with

Developing Knowledge	On-level Knowledge	Advanced Knowledge	English Language Development
☐ Cooperative Learning TE p. 35	☐ Cooperative Learning TE p. 35	☐ Cooperative Learning TE p. 35	☐ Cooperative Learning TE p. 35
☐ Practice A 1-7 CRB	☐ Practice B 1-7 CRB	☐ Practice C 1-7 CRB	☐ Practice A, B, or C 1-7 CRB
☐ Reteach 1-7 CRB	☐ Puzzles, Twisters & Teasers 1-7 CRB	☐ Challenge 1-7 CRB	☐ *Success for ELL* 1-7
☐ Homework Help Online Keyword: MT7 1-7	☐ Homework Help Online Keyword: MT7 1-7	☐ Homework Help Online Keyword: MT7 1-7	☐ Homework Help Online Keyword: MT7 1-7
☐ *Lesson Tutorial Video* 1-7	☐ *Lesson Tutorial Video* 1-7	☐ *Lesson Tutorial Video* 1-7	☐ *Lesson Tutorial Video* 1-7
☐ Reading Strategies 1-7 CRB	☐ Problem Solving 1-7 CRB	☐ Problem Solving 1-7 CRB	☐ Reading Strategies 1-7 CRB
☐ *Questioning Strategies* pp. 12–13			☐ Lesson Vocabulary SE p. 34
☐ *IDEA Works!* 1-7			☐ *Multilingual Glossary*

ASSESSMENT
☐ Lesson Quiz, TE p. 38 and DT 1-7 ☐ State-Specific Test Prep Online Keyword: MT7 TestPrep

Teacher's Name _____ Class _____ Date _____

Lesson Plan 1-8
Solving Equations by Multiplying or Dividing pp. 39–43   Day _____

Objective Students solve equations using multiplication and division.

> **NCTM Standards:** Understand patterns, relations, and functions; Recognize reasoning and proof as fundamental aspects of mathematics; Make and investigate mathematical conjectures; Select and use various types of reasoning and methods of proof.

Pacing
- [ ] 45-minute Classes: 1 day
- [ ] 90-minute Classes: 1/2 day
- [ ] Other _____

WARM UP
- [ ] Warm Up TE p. 39 and Daily Transparency 1-8
- [ ] Problem of the Day TE p. 39 and Daily Transparency 1-8
- [ ] Countdown to Testing Transparency Week 2

TEACH
- [ ] Lesson Presentation CD-ROM 1-8
- [ ] Alternate Opener, Explorations Transparency 1-8, TE p. 39, and Exploration 1-8
- [ ] Reaching All Learners TE p. 40
- [ ] Teaching Transparency 1-8
- [ ] *Hands-On Lab Activities* 1-8
- [ ] *Know-It Notebook* 1-8

PRACTICE AND APPLY
- [ ] Example 1: Average: 1–8, 22–29, 68–77 Advanced: 22–29, 43–46, 67–77
- [ ] Example 2: Average: 1–16, 22–37, 60, 68–77 Advanced: 22–37, 43–50, 66–77
- [ ] Example 3: Average: 1–17, 23–37 odd, 60–63, 68–77 Advanced: 22–38, 59–77
- [ ] Example 4: Average: 1–41 odd, 59–63, 68–77 Advanced: 23–57 odd, 59–77

REACHING ALL LEARNERS – Differentiated Instruction for students with

Developing Knowledge	On-level Knowledge	Advanced Knowledge	English Language Development
☐ Inclusion TE p. 40	☐ Concrete Manipulatives TE p. 40	☐ Concrete Manipulatives TE p. 40	☐ Concrete Manipulatives TE p. 40
☐ Practice A 1-8 CRB	☐ Practice B 1-8 CRB	☐ Practice C 1-8 CRB	☐ Practice A, B, or C 1-8 CRB
☐ Reteach 1-8 CRB	☐ Puzzles, Twisters & Teasers 1-8 CRB	☐ Challenge 1-8 CRB	☐ *Success for ELL* 1-8
☐ Homework Help Online Keyword: MT7 1-8	☐ Homework Help Online Keyword: MT7 1-8	☐ Homework Help Online Keyword: MT7 1-8	☐ Homework Help Online Keyword: MT7 1-8
☐ *Lesson Tutorial Video* 1-8	☐ *Lesson Tutorial Video* 1-8	☐ *Lesson Tutorial Video* 1-8	☐ *Lesson Tutorial Video* 1-8
☐ Reading Strategies 1-8 CRB	☐ Problem Solving 1-8 CRB	☐ Problem Solving 1-8 CRB	☐ Reading Strategies 1-8 CRB
☐ *Questioning Strategies* pp. 14–15			
☐ *IDEA Works!* 1-8			☐ *Multilingual Glossary*

ASSESSMENT
- [ ] Lesson Quiz, TE p. 43 and DT 1-8
- [ ] State-Specific Test Prep Online Keyword: MT7 TestPrep

Teacher's Name _____ Class _____ Date _____

Lesson Plan 1-9
Introduction to Inequalities pp. 44–47   Day _____

Objective Students solve and graph inequalities.

> **NCTM Standards:** Understand meanings of operations and how they relate to one another.

Pacing
- ☐ 45-minute Classes: 1 day   ☐ 90-minute Classes: 1/2 day   ☐ Other_____

WARM UP
- ☐ Warm Up TE p. 44 and Daily Transparency 1-9
- ☐ Problem of the Day TE p. 44 and Daily Transparency 1-9
- ☐ Countdown to Testing Transparency Week 2

TEACH
- ☐ Lesson Presentation CD-ROM 1-9
- ☐ Alternate Opener, Explorations Transparency 1-9, TE p. 44, and Exploration 1-9
- ☐ Reaching All Learners TE p. 45
- ☐ Teaching Transparency 1-9
- ☐ *Know-It Notebook* 1-9

PRACTICE AND APPLY
- ☐ Example 1: Average: 1–6, 15–20, 38, 56–65 Advanced: 15–20, 54–65
- ☐ Example 2: Average: 1–43 odd, 56–65 Advanced: 15–28, 45–65

REACHING ALL LEARNERS – Differentiated Instruction for students with

Developing Knowledge	On-level Knowledge	Advanced Knowledge	English Language Development
☐ Cooperative Learning TE p. 45	☐ Cooperative Learning TE p. 45	☐ Cooperative Learning TE p. 45	☐ Cooperative Learning TE p. 45
☐ Practice A 1-9 CRB	☐ Practice B 1-9 CRB	☐ Practice C 1-9 CRB	☐ Practice A, B, or C 1-9 CRB
☐ Reteach 1-9 CRB	☐ Puzzles, Twisters & Teasers 1-9 CRB	☐ Challenge 1-9 CRB	☐ *Success for ELL* 1-9
☐ Homework Help Online Keyword: MT7 1-9	☐ Homework Help Online Keyword: MT7 1-9	☐ Homework Help Online Keyword: MT7 1-9	☐ Homework Help Online Keyword: MT7 1-9
☐ *Lesson Tutorial Video* 1-9	☐ *Lesson Tutorial Video* 1-9	☐ *Lesson Tutorial Video* 1-9	☐ *Lesson Tutorial Video* 1-9
☐ Reading Strategies 1-9 CRB	☐ Problem Solving 1-9 CRB	☐ Problem Solving 1-9 CRB	☐ Reading Strategies 1-9 CRB
☐ *Questioning Strategies* pp. 16–17	☐ Cognitive Strategies TE p. 45	☐ Cognitive Strategies TE p. 45	☐ Lesson Vocabulary SE p. 44
☐ *IDEA Works!* 1-9			☐ *Multilingual Glossary*

ASSESSMENT
- ☐ Lesson Quiz, TE p. 47 and DT 1-9   ☐ State-Specific Test Prep Online Keyword: MT7 TestPrep

Teacher's Name _____ Class _____ Date _____

Lesson Plan 2-1
Rational Numbers pp. 64–67 Day _____

Objective Students write rational numbers in equivalent forms.

NCTM Standards: Understand numbers, ways of representing numbers, relationships among numbers, and number systems.

Pacing
- ☐ 45-minute Classes: 1 day   ☐ 90-minute Classes: 1/2 day   ☐ Other _____

WARM UP
- ☐ Warm Up TE p. 64 and Daily Transparency 2-1
- ☐ Problem of the Day TE p. 64 and Daily Transparency 2-1
- ☐ Countdown to Testing Transparency Week 3

TEACH
- ☐ Lesson Presentation CD-ROM 2-1
- ☐ Alternate Opener, Explorations Transparency 2-1, TE p. 64, and Exploration 2-1
- ☐ Reaching All Learners TE p. 65
- ☐ Teaching Transparency 2-1
- ☐ *Technology Lab Activities* 2-1
- ☐ *Know-It Notebook* 2-1

PRACTICE AND APPLY
- ☐ Example 1: Average: 1–10, 29–38, 67–79 Advanced: 29–38, 57, 58, 63, 64, 66–79
- ☐ Example 2: Average: 1–18, 29–46, 59, 67–79 Advanced: 29–46, 57–59, 63, 64, 66–79
- ☐ Example 3: Average: 1–60, 67–79 Advanced: 34–38, 43–51, 57–79

REACHING ALL LEARNERS – Differentiated Instruction for students with

Developing Knowledge	On-level Knowledge	Advanced Knowledge	English Language Development
☐ Multiple Representations TE p. 65	☐ Multiple Representations TE p. 65	☐ Multiple Representations TE p. 65	☐ Multiple Representations TE p. 65
☐ Practice A 2-1 CRB	☐ Practice B 2-1 CRB	☐ Practice C 2-1 CRB	☐ Practice A, B, or C 2-1 CRB
☐ Reteach 2-1 CRB	☐ Puzzles, Twisters & Teasers 2-1 CRB	☐ Challenge 2-1 CRB	☐ *Success for ELL* 2-1
☐ Homework Help Online Keyword: MT7 2-1	☐ Homework Help Online Keyword: MT7 2-1	☐ Homework Help Online Keyword: MT7 2-1	☐ Homework Help Online Keyword: MT7 2-1
☐ *Lesson Tutorial Video* 2-1	☐ *Lesson Tutorial Video* 2-1	☐ *Lesson Tutorial Video* 2-1	☐ *Lesson Tutorial Video* 2-1
☐ Reading Strategies 2-1 CRB	☐ Problem Solving 2-1 CRB	☐ Problem Solving 2-1 CRB	☐ Reading Strategies 2-1 CRB
☐ *Questioning Strategies* pp. 18–19			☐ Lesson Vocabulary p. 64
☐ *IDEA Works!* 2-1			☐ *Multilingual Glossary*

ASSESSMENT
- ☐ Lesson Quiz, TE p. 67 and DT 2-1   ☐ State-Specific Test Prep Online Keyword: MT7 TestPrep

Teacher's Name _____ Class _____ Date _____

Lesson Plan 2-2
Comparing and Ordering Rational Numbers pp. 68–71   Day _____

Objective Students compare and order positive and negative rational numbers written as fractions, decimals, and integers.

> **NCTM Standards:** Represent and analyze mathematical situations and structures using algebraic symbols; Recognize reasoning and proof as fundamental aspects of mathematics; Create and use representations to organize, record, and communicate mathematical ideas.

Pacing
☐ 45-minute Classes: 1 day   ☐ 90-minute Classes: 1/2 day   ☐ Other _____

WARM UP
☐ Warm Up TE p. 68 and Daily Transparency 2-2
☐ Problem of the Day TE p. 68 and Daily Transparency 2-2
☐ Countdown to Testing Transparency Week 3

TEACH
☐ Lesson Presentation CD-ROM 2-2
☐ Alternate Opener, Explorations Transparency 2-2, TE p. 68, and Exploration 2-2
☐ Reaching All Learners TE p. 69
☐ *Know-It Notebook* 2-2

PRACTICE AND APPLY
☐ Example 1: Average: 1–4, 10–17, 27, 42–53 Advanced: 10–17, 27, 31, 42–53
☐ Example 2: Average: 1–8, 10–25, 27–30, 42–53 Advanced: 10–25, 27–34, 39–53
☐ Example 3: Average: 1–30, 37, 38, 42–53 Advanced: 10–24, 22–53

REACHING ALL LEARNERS – Differentiated Instruction for students with

Developing Knowledge	On-level Knowledge	Advanced Knowledge	English Language Development
☐ Concrete Manipulatives TE p. 69	☐ Concrete Manipulatives TE p. 69	☐ Concrete Manipulatives TE p. 69	☐ Concrete Manipulatives TE p. 69
☐ Practice A 2-2 CRB	☐ Practice B 2-2 CRB	☐ Practice C 2-2 CRB	☐ Practice A, B, or C 2-2 CRB
☐ Reteach 2-2 CRB	☐ Puzzles, Twisters & Teasers 2-2 CRB	☐ Challenge 2-2 CRB	☐ *Success for ELL* 2-2
☐ Homework Help Online Keyword: MT7 2-2	☐ Homework Help Online Keyword: MT7 2-2	☐ Homework Help Online Keyword: MT7 2-2	☐ Homework Help Online Keyword: MT7 2-2
☐ *Lesson Tutorial Video* 2-2	☐ *Lesson Tutorial Video* 2-2	☐ *Lesson Tutorial Video* 2-2	☐ *Lesson Tutorial Video* 2-2
☐ Reading Strategies 2-2 CRB	☐ Problem Solving 2-2 CRB	☐ Problem Solving 2-2 CRB	☐ Reading Strategies 2-2 CRB
☐ *Questioning Strategies* pp. 20–21			☐ Lesson Vocabulary p. 68
☐ *IDEA Works!* 2-2			☐ *Multilingual Glossary*

ASSESSMENT
☐ Lesson Quiz, TE p. 71 and DT 2-2   ☐ State-Specific Test Prep Online Keyword: MT7 TestPrep

Teacher's Name _____ Class _____ Date _____

Lesson Plan 2-3
Adding and Subtracting Rational Numbers pp. 72–75 Day _____

Objective Students add and subtract decimals and rational numbers with like denominators.

> **NCTM Standards:** Understand meanings of operations and how they relate to one another; Compute fluently and make reasonable estimates; Recognize and apply mathematics in contexts outside of mathematics.

Pacing
☐ 45-minute Classes: 1 day ☐ 90-minute Classes: 1/2 day ☐ Other_____

WARM UP
☐ Warm Up TE p. 72 and Daily Transparency 2-3
☐ Problem of the Day TE p. 72 and Daily Transparency 2-3
☐ Countdown to Testing Transparency Week 3

TEACH
☐ Lesson Presentation CD-ROM 2-3
☐ Alternate Opener, Explorations Transparency 2-3, TE p. 72, and Exploration 2-3
☐ Reaching All Learners TE p. 73
☐ Teaching Transparency 2-3
☐ *Hands-On Lab Activities* 2-3
☐ *Know-It Notebook* 2-3

PRACTICE AND APPLY
☐ Example 1: Average: 1, 13, 43–53 Advanced: 13, 38, 39, 43–53
☐ Example 2: Average: 1–5, 13–17, 43–53 Advanced: 13–17, 32–34, 43–53
☐ Example 3: Average: 1–9, 13–21, 29, 30–33, 43–53 Advanced: 13–21, 28, 29, 33–36, 38–41, 43–53
☐ Example 4: Average: 1–25, 29, 30–33, 43–53 Advanced: 13–28, 29, 36–53

REACHING ALL LEARNERS – Differentiated Instruction for students with

Developing Knowledge	On-level Knowledge	Advanced Knowledge	English Language Development
☐ Critical Thinking TE p. 73	☐ Critical Thinking TE p. 73	☐ Critical Thinking TE p. 73	☐ Critical Thinking TE p. 73
☐ Practice A 2-3 CRB	☐ Practice B 2-3 CRB	☐ Practice C 2-3 CRB	☐ Practice A, B, or C 2-3 CRB
☐ Reteach 2-3 CRB	☐ Puzzles, Twisters & Teasers 2-3 CRB	☐ Challenge 2-3 CRB	☐ *Success for ELL* 2-3
☐ Homework Help Online Keyword: MT7 2-3	☐ Homework Help Online Keyword: MT7 2-3	☐ Homework Help Online Keyword: MT7 2-3	☐ Homework Help Online Keyword: MT7 2-3
☐ *Lesson Tutorial Video* 2-3	☐ *Lesson Tutorial Video* 2-3	☐ *Lesson Tutorial Video* 2-3	☐ *Lesson Tutorial Video* 2-3
☐ Reading Strategies 2-3 CRB	☐ Problem Solving 2-3 CRB	☐ Problem Solving 2-3 CRB	☐ Reading Strategies 2-3 CRB
☐ Questioning Strategies pp. 22–23			
☐ *IDEA Works!* 2-3			☐ *Multilingual Glossary*

ASSESSMENT
☐ Lesson Quiz, TE p. 75 and DT 2-3 ☐ State-Specific Test Prep Online Keyword: MT7 TestPrep

Teacher's Name _____ Class _____ Date _____

Lesson Plan 2-4
Multiplying Rational Numbers pp. 76–79   Day _____

Objective Students multiply fractions, mixed numbers, and decimals.

> **NCTM Standards:** Compute fluently and make reasonable estimates; Recognize and apply mathematics in contexts outside of mathematics.

Pacing
☐ 45-minute Classes: 1 day   ☐ 90-minute Classes: 1/2 day   ☐ Other_____

WARM UP
☐ Warm Up TE p. 76 and Daily Transparency 2-4
☐ Problem of the Day TE p. 76 and Daily Transparency 2-4
☐ Countdown to Testing Transparency Week 3

TEACH
☐ Lesson Presentation CD-ROM 2-4
☐ Alternate Opener, Explorations Transparency 2-4, TE p. 76, and Exploration 2-4
☐ Reaching All Learners TE p. 77
☐ Teaching Transparency 2-4
☐ *Know-It Notebook* 2-4

PRACTICE AND APPLY
☐ Example 1: Average: 1–4, 14–21, 42, 43, 57–67 Advanced: 14–21, 40–43, 57–67
☐ Example 2: Average: 1–8, 14–29, 42, 43, 48, 49, 57–67 Advanced: 14–29, 41, 42, 48–51, 54, 56–67
☐ Example 3: Average: 1–12, 14–37, 42, 43, 46–49, 57–67 Advanced: 14–37 even, 41–51 odd, 54, 56–67
☐ Example 4: Average: 1–39, 42–52, 57–67 Advanced: 14–52 even, 52–67

REACHING ALL LEARNERS – Differentiated Instruction for students with

Developing Knowledge	On-level Knowledge	Advanced Knowledge	English Language Development
☐ Multiple Representations TE p. 77	☐ Multiple Representations TE p. 77	☐ Multiple Representations TE p. 77	☐ Multiple Representations TE p. 77
☐ Practice A 2-4 CRB	☐ Practice B 2-4 CRB	☐ Practice C 2-4 CRB	☐ Practice A, B, or C 2-4 CRB
☐ Reteach 2-4 CRB	☐ Puzzles, Twisters & Teasers 2-4 CRB	☐ Challenge 2-4 CRB	☐ *Success for ELL* 2-4
☐ Homework Help Online Keyword: MT7 2-4	☐ Homework Help Online Keyword: MT7 2-4	☐ Homework Help Online Keyword: MT7 2-4	☐ Homework Help Online Keyword: MT7 2-4
☐ *Lesson Tutorial Video* 2-4	☐ *Lesson Tutorial Video* 2-4	☐ *Lesson Tutorial Video* 2-4	☐ *Lesson Tutorial Video* 2-4
☐ Reading Strategies 2-4 CRB	☐ Problem Solving 2-4 CRB	☐ Problem Solving 2-4 CRB	☐ Reading Strategies 2-4 CRB
☐ *Questioning Strategies* pp. 24–25	☐ Number Sense TE p. 77	☐ Number Sense TE p. 77	
☐ *IDEA Works!* 2-4			☐ *Multilingual Glossary*

ASSESSMENT
☐ Lesson Quiz, TE p. 79 and DT 2-4   ☐ State-Specific Test Prep Online Keyword: MT7 TestPrep

Teacher's Name _____ Class _____ Date _____

Lesson Plan 2-5
Dividing Rational Numbers pp. 80–84   Day _____

Objective Students divide fractions and decimals.

> **NCTM Standards:** Understand meanings of operations and how they relate to one another; Compute fluently and make reasonable estimates.

Pacing
☐ 45-minute Classes: 1 day ☐ 90-minute Classes: 1/2 day ☐ Other _____

WARM UP
☐ Warm Up TE p. 80 and Daily Transparency 2-5
☐ Problem of the Day TE p. 80 and Daily Transparency 2-5
☐ Countdown to Testing Transparency Week 3

TEACH
☐ Lesson Presentation CD-ROM 2-5
☐ Alternate Opener, Explorations Transparency 2-5, TE p. 80, and Exploration 2-5
☐ Reaching All Learners TE p. 81
☐ Teaching Transparency 2-5
☐ *Hands-On Lab Activities* 2-5
☐ *Know-It Notebook* 2-5

PRACTICE AND APPLY
☐ Example 1: Average: 1–8, 24–31, 54–63 Advanced: 24–31, 52, 54–63
☐ Example 2: Average: 1–16, 24–39, 54–63 Advanced: 24–39, 52, 54–63
☐ Example 3: Average: 1–22, 24–45, 54–63 Advanced: 24–45, 52, 54–63
☐ Example 4: Average: 1–46, 47–49, 54–63 Advanced: 24–46 even, 47–63

REACHING ALL LEARNERS – Differentiated Instruction for students with

Developing Knowledge	On-level Knowledge	Advanced Knowledge	English Language Development
☐ Inclusion TE p. 82	☐ Cognitive Strategies TE p. 81	☐ Cognitive Strategies TE p. 81	☐ Cognitive Strategies TE p. 81
☐ Practice A 2-5 CRB	☐ Practice B 2-5 CRB	☐ Practice C 2-5 CRB	☐ Practice A, B, or C 2-5 CRB
☐ Reteach 2-5 CRB	☐ Puzzles, Twisters & Teasers 2-5 CRB	☐ Challenge 2-5 CRB	☐ *Success for ELL* 2-5
☐ Homework Help Online Keyword: MT7 2-5	☐ Homework Help Online Keyword: MT7 2-5	☐ Homework Help Online Keyword: MT7 2-5	☐ Homework Help Online Keyword: MT7 2-5
☐ *Lesson Tutorial Video* 2-5	☐ *Lesson Tutorial Video* 2-5	☐ *Lesson Tutorial Video* 2-5	☐ *Lesson Tutorial Video* 2-5
☐ Reading Strategies 2-5 CRB	☐ Problem Solving 2-5 CRB	☐ Problem Solving 2-5 CRB	☐ Reading Strategies 2-5 CRB
☐ *Questioning Strategies* pp. 26–27	☐ Number Sense TE p. 81	☐ Number Sense TE p. 81	☐ Lesson Vocabulary SE p. 80
☐ *IDEA Works!* 2-5			☐ *Multilingual Glossary*

ASSESSMENT
☐ Lesson Quiz, TE p. 84 and DT 2-5   ☐ State-Specific Test Prep Online Keyword: MT7 TestPrep

Copyright © Holt, Rinehart and Winston.
All rights reserved.

Holt Mathematics

Teacher's Name _____ Class _____ Date _____

Lesson Plan 2-6
Adding and Subtracting with Unlike Denominators pp. 85–88   Day _____

Objective Students add and subtract fractions with unlike denominators.

> **NCTM Standards:** Understand meanings of operations and how they relate to one another; Compute fluently and make reasonable estimates.

Pacing
☐ 45-minute Classes: 1 day   ☐ 90-minute Classes: 1/2 day   ☐ Other_____

WARM UP
☐ Warm Up TE p. 85 and Daily Transparency 2-6
☐ Problem of the Day TE p. 85 and Daily Transparency 2-6
☐ Countdown to Testing Transparency Week 4

TEACH
☐ Lesson Presentation CD-ROM 2-6
☐ Alternate Opener, Explorations Transparency 2-6, TE p. 85, and Exploration 2-6
☐ Reaching All Learners TE p. 86
☐ *Know-It Notebook* 2-6

PRACTICE AND APPLY
☐ Example 1: Average: 1–4, 9–16, 32–40 Advanced: 9–16, 25, 32–40
☐ Example 2: Average: 1–7, 9–22, 32–40 Advanced: 16–22, 25, 32–40
☐ Example 3: Average: 1–23, 25, 28, 29, 32–40 Advanced: 13–20, 24–40

REACHING ALL LEARNERS – Differentiated Instruction for students with

Developing Knowledge	On-level Knowledge	Advanced Knowledge	English Language Development
☐ Concrete Manipulatives TE p. 86	☐ Concrete Manipulatives TE p. 86	☐ Concrete Manipulatives TE p. 86	☐ Concrete Manipulatives TE p. 86
☐ Practice A 2-6 CRB	☐ Practice B 2-6 CRB	☐ Practice C 2-6 CRB	☐ Practice A, B, or C 2-6 CRB
☐ Reteach 2-6 CRB	☐ Puzzles, Twisters & Teasers 2-6 CRB	☐ Challenge 2-6 CRB	☐ *Success for ELL* 2-6
☐ Homework Help Online Keyword: MT7 2-6	☐ Homework Help Online Keyword: MT7 2-6	☐ Homework Help Online Keyword: MT7 2-6	☐ Homework Help Online Keyword: MT7 2-6
☐ *Lesson Tutorial Video* 2-6	☐ *Lesson Tutorial Video* 2-6	☐ *Lesson Tutorial Video* 2-6	☐ *Lesson Tutorial Video* 2-6
☐ Reading Strategies 2-6 CRB	☐ Problem Solving 2-6 CRB	☐ Problem Solving 2-6 CRB	☐ Reading Strategies 2-6 CRB
☐ *Questioning Strategies* pp. 28–29			
☐ *IDEA Works!* 2-6			☐ *Multilingual Glossary*

ASSESSMENT
☐ Lesson Quiz, TE p. 88 and DT 2-6   ☐ State-Specific Test Prep Online Keyword: MT7 TestPrep

Teacher's Name _____ Class _____ Date _____

Lesson Plan 2-7
Solving Equations with Rational Numbers pp. 92–95   Day _____

Objective Students solve equations with rational numbers.

> **NCTM Standards:** Understand meanings of operations and how they relate to one another; Recognize and apply mathematics in contexts outside of mathematics.

Pacing
☐ 45-minute Classes: 1 day   ☐ 90-minute Classes: 1/2 day   ☐ Other_____

WARM UP
☐ Warm Up TE p. 92 and Daily Transparency 2-7
☐ Problem of the Day TE p. 92 and Daily Transparency 2-7
☐ Countdown to Testing Transparency Week 4

TEACH
☐ Lesson Presentation CD-ROM 2-7
☐ Alternate Opener, Explorations Transparency 2-7, TE p. 92, and Exploration 2-7
☐ Reaching All Learners TE p. 93
☐ *Know-It Notebook* 2-7

PRACTICE AND APPLY
☐ Example 1: Average: 1–6, 14–19, 34, 35, 54–61   Advanced: 14–19, 47–49, 54–61
☐ Example 2: Average: 1–12, 14–27, 33–40, 54–61   Advanced: 17–23, 37–46, 52, 54–61
☐ Example 3: Average: 1–40, 54–61   Advanced: 17–23, 29–43, 50–61

REACHING ALL LEARNERS – Differentiated Instruction for students with

Developing Knowledge	On-level Knowledge	Advanced Knowledge	English Language Development
☐ Cooperative Learning TE p. 93	☐ Cooperative Learning TE p. 93	☐ Cooperative Learning TE p. 93	☐ Cooperative Learning TE p. 93
☐ Practice A 2-7 CRB	☐ Practice B 2-7 CRB	☐ Practice C 2-7 CRB	☐ Practice A, B, or C 2-7 CRB
☐ Reteach 2-7 CRB	☐ Puzzles, Twisters & Teasers 2-7 CRB	☐ Challenge 2-7 CRB	☐ *Success for ELL* 2-7
☐ Homework Help Online Keyword: MT7 2-7	☐ Homework Help Online Keyword: MT7 2-7	☐ Homework Help Online Keyword: MT7 2-7	☐ Homework Help Online Keyword: MT7 2-7
☐ *Lesson Tutorial Video* 2-7	☐ *Lesson Tutorial Video* 2-7	☐ *Lesson Tutorial Video* 2-7	☐ *Lesson Tutorial Video* 2-7
☐ Reading Strategies 2-7 CRB	☐ Problem Solving 2-7 CRB	☐ Problem Solving 2-7 CRB	☐ Reading Strategies 2-7 CRB
☐ *Questioning Strategies* pp. 30–31			
☐ *IDEA Works!* 2-7			☐ *Multilingual Glossary*

ASSESSMENT
☐ Lesson Quiz, TE p. 95 and DT 2-7   ☐ State-Specific Test Prep Online Keyword: MT7 TestPrep

Teacher's Name _____ Class _____ Date _____

Lesson Plan 2-8
Solving Two-Step Equations pp. 98–101   Day _____

Objective Students solve two-step equations.

> **NCTM Standards:** Understand meanings of operations and how they relate to one another; Use mathematical models to represent and understand quantitative relationships; Understand how mathematical ideas interconnect and build on one another to produce a coherent whole.

Pacing
☐ 45-minute Classes: 1 day   ☐ 90-minute Classes: 1/2 day   ☐ Other_____

WARM UP
☐ Warm Up TE p. 98 and Daily Transparency 2-8
☐ Problem of the Day TE p. 98 and Daily Transparency 2-8
☐ Countdown to Testing Transparency Week 4

TEACH
☐ Lesson Presentation CD-ROM 2-8
☐ Alternate Opener, Explorations Transparency 2-8, TE p. 98, and Exploration 2-8
☐ Reaching All Learners TE p. 99
☐ *Hands-On Lab Activities* 2-8
☐ *Know-It Notebook* 2-8

PRACTICE AND APPLY
☐ Example 1: Average: 1, 10, 34–36, 38–41, 43–52 Advanced: 10, 34–52
☐ Example 2: Average: 1–18, 19–26, 34–36, 38–41, 43–52 Advanced: 10, 25–52

REACHING ALL LEARNERS – Differentiated Instruction for students with

Developing Knowledge	On-level Knowledge	Advanced Knowledge	English Language Development
☐ Inclusion TE p. 99	☐ Cooperative Learning TE p. 99	☐ Cooperative Learning TE p. 99	☐ Cooperative Learning TE p. 99
☐ Practice A 2-8 CRB	☐ Practice B 2-8 CRB	☐ Practice C 2-8 CRB	☐ Practice A, B, or C 2-8 CRB
☐ Reteach 2-8 CRB	☐ Puzzles, Twisters & Teasers 2-8 CRB	☐ Challenge 2-8 CRB	☐ *Success for ELL* 2-8
☐ Homework Help Online Keyword: MT7 2-8	☐ Homework Help Online Keyword: MT7 2-8	☐ Homework Help Online Keyword: MT7 2-8	☐ Homework Help Online Keyword: MT7 2-8
☐ *Lesson Tutorial Video* 2-8	☐ *Lesson Tutorial Video* 2-8	☐ *Lesson Tutorial Video* 2-8	☐ *Lesson Tutorial Video* 2-8
☐ Reading Strategies 2-8 CRB	☐ Problem Solving 2-8 CRB	☐ Problem Solving 2-8 CRB	☐ Reading Strategies 2-8 CRB
☐ *Questioning Strategies* pp. 32–33			
☐ *IDEA Works!* 2-8			☐ *Multilingual Glossary*

ASSESSMENT
☐ Lesson Quiz, TE p. 101 and DT 2-8   ☐ State-Specific Test Prep Online Keyword: MT7 TestPrep

Teacher's Name _____ Class _____ Date _____

Lesson Plan 3-1
Ordered Pairs pp. 118–121 Day _____

Objective Students write solutions of equations in two variables as ordered pairs.

> **NCTM Standards:** Understand patterns, relations, and functions; Specify locations and describe spatial relationships using coordinate geometry and other representational systems; Create and use representations to organize, record, and communicate mathematical ideas.

Pacing
☐ 45-minute Classes: 1 day ☐ 90-minute Classes: 1/2 day ☐ Other_____

WARM UP
☐ Warm Up TE p. 118 and Daily Transparency 3-1
☐ Problem of the Day TE p. 118 and Daily Transparency 3-1
☐ Countdown to Testing Transparency Week 5

TEACH
☐ Lesson Presentation CD-ROM 3-1
☐ Alternate Opener, Explorations Transparency 3-1, TE p. 118, and Exploration 3-1
☐ Reaching All Learners TE p. 119
☐ *Technology Lab Activities* 3-1
☐ *Know-It Notebook* 3-1

PRACTICE AND APPLY
☐ Example 1: Average: 1–4, 8–11, 23, 24, 36–44 Advanced: 8–11, 17–24, 32, 36–44
☐ Example 2: Average: 1–6, 8–15, 23, 24, 27, 28, 36–44 Advanced: 8–11, 17–24, 29, 30, 32–34, 36–44
☐ Example 3: Average: 1–16, 23, 24, 26–28, 31, 36–44 Advanced: 8–26, 29–44

REACHING ALL LEARNERS – Differentiated Instruction for students with

Developing Knowledge	On-level Knowledge	Advanced Knowledge	English Language Development
☐ Visual Cues TE p. 119	☐ Visual Cues TE p. 119	☐ Visual Cues TE p. 119	☐ Visual Cues TE p. 119
☐ Practice A 3-1 CRB	☐ Practice B 3-1 CRB	☐ Practice C 3-1 CRB	☐ Practice A, B, or C 3-1 CRB
☐ Reteach 3-1 CRB	☐ Puzzles, Twisters & Teasers 3-1 CRB	☐ Challenge 3-1 CRB	☐ *Success for ELL* 3-1
☐ Homework Help Online Keyword: MT7 3-1	☐ Homework Help Online Keyword: MT7 3-1	☐ Homework Help Online Keyword: MT7 3-1	☐ Homework Help Online Keyword: MT7 3-1
☐ *Lesson Tutorial Video* 3-1	☐ *Lesson Tutorial Video* 3-1	☐ *Lesson Tutorial Video* 3-1	☐ *Lesson Tutorial Video* 3-1
☐ Reading Strategies 3-1 CRB	☐ Problem Solving 3-1 CRB	☐ Problem Solving 3-1 CRB	☐ Reading Strategies 3-1 CRB
☐ *Questioning Strategies* pp. 34–35			☐ Lesson Vocabulary SE p. 118
☐ *IDEA Works!* 3-1			☐ *Multilingual Glossary*

ASSESSMENT
☐ Lesson Quiz, TE p. 121 and DT 3-1 ☐ State-Specific Test Prep Online Keyword: MT7 TestPrep

Teacher's Name _____ Class _____ Date _____

Lesson Plan 3-2
Graphing on a Coordinate Plane pp. *122–125*   Day _____

Objective Students graph points and lines on the coordinate plane.

> **NCTM Standards:** Understand patterns, relations, and functions; Specify locations and describe spatial relationships using coordinate geometry and other representational systems.

Pacing
☐ 45-minute Classes: 1 day   ☐ 90-minute Classes: 1/2 day   ☐ Other_____

WARM UP
☐ Warm Up TE p. 122 and Daily Transparency 3-2
☐ Problem of the Day TE p. 122 and Daily Transparency 3-2
☐ Countdown to Testing Transparency Week 5

TEACH
☐ Lesson Presentation CD-ROM 3-2
☐ Alternate Opener, Explorations Transparency 3-2, TE p. 122, and Exploration 3-2
☐ Reaching All Learners TE p. 123
☐ Teaching Transparency 3-2
☐ *Know-It Notebook* 3-2

PRACTICE AND APPLY
☐ Example 1: Average: 1–6, 17–22, 38–47 Advanced: 17–22, 36, 39–48
☐ Example 2: Average: 1–14, 17–30, 38–47 Advanced: 17–30, 36, 38–47
☐ Example 3: Average: 1–32, 34, 38–47 Advanced: 17–47

REACHING ALL LEARNERS – Differentiated Instruction for students with

Developing Knowledge	On-level Knowledge	Advanced Knowledge	English Language Development
☐ Critical Thinking TE p. 123	☐ Critical Thinking TE p. 123	☐ Critical Thinking TE p. 123	☐ Critical Thinking TE p. 123
☐ Practice A 3-2 CRB	☐ Practice B 3-2 CRB	☐ Practice C 3-2 CRB	☐ Practice A, B, or C 3-2 CRB
☐ Reteach 3-2 CRB	☐ Puzzles, Twisters & Teasers 3-2 CRB	☐ Challenge 3-2 CRB	☐ *Success for ELL* 3-2
☐ Homework Help Online Keyword: MT7 3-2	☐ Homework Help Online Keyword: MT7 3-2	☐ Homework Help Online Keyword: MT7 3-2	☐ Homework Help Online Keyword: MT7 3-2
☐ *Lesson Tutorial Video* 3-2	☐ *Lesson Tutorial Video* 3-2	☐ *Lesson Tutorial Video* 3-2	☐ *Lesson Tutorial Video* 3-2
☐ Reading Strategies 3-2 CRB	☐ Problem Solving 3-2 CRB	☐ Problem Solving 3-2 CRB	☐ Reading Strategies 3-2 CRB
☐ *Questioning Strategies* pp. 36–37			☐ Lesson Vocabulary SE p. 122
☐ *IDEA Works!* 3-2			☐ *Multilingual Glossary*

ASSESSMENT
☐ Lesson Quiz, TE p. 125 and DT 3-2   ☐ State-Specific Test Prep Online Keyword: MT7 TestPrep

Teacher's Name _____ Class _____ Date _____

Lesson Plan 3-3
Interpreting Graphs and Tables pp. 127–131   Day _____

Objective Students interpret information given in a graph or a table and make a graph to solve problems.

> **NCTM Standards:** Understand patterns, relations, and functions; Create and use representations to organize, record, and communicate mathematical ideas; Select, apply, and translate among mathematical representations to solve problems.

Pacing
☐ 45-minute Classes: 1 day   ☐ 90-minute Classes: 1/2 day   ☐ Other_____

WARM UP
☐ Warm Up TE p. 127 and Daily Transparency 3-3
☐ Problem of the Day TE p. 127 and Daily Transparency 3-3
☐ Countdown to Testing Transparency Week 5

TEACH
☐ Lesson Presentation CD-ROM 3-3
☐ Alternate Opener, Explorations Transparency 3-3, TE p. 127, and Exploration 3-3
☐ Reaching All Learners TE p. 128
☐ *Technology Lab Activities* 3-3
☐ *Know-It Notebook* 3-3

PRACTICE AND APPLY
☐ Example 1: Average: 1, 5, 12–21 Advanced: 5, 12–21
☐ Example 2: Average: 1, 2, 5, 6, 10, 12–21 Advanced: 5, 6, 10, 12–21
☐ Example 3: Average: 1–7, 10, 12–21 Advanced: 5–21

REACHING ALL LEARNERS – Differentiated Instruction for students with

Developing Knowledge	On-level Knowledge	Advanced Knowledge	English Language Development
☐ Kinesthetic Experience TE p. 128	☐ Kinesthetic Experience TE p. 128	☐ Kinesthetic Experience TE p. 128	☐ Kinesthetic Experience TE p. 128
☐ Practice A 3-3 CRB	☐ Practice B 3-3 CRB	☐ Practice C 3-3 CRB	☐ Practice A, B, or C 3-3 CRB
☐ Reteach 3-3 CRB	☐ Puzzles, Twisters & Teasers 3-3 CRB	☐ Challenge 3-3 CRB	☐ *Success for ELL* 3-3
☐ Homework Help Online Keyword: MT7 3-3	☐ Homework Help Online Keyword: MT7 3-3	☐ Homework Help Online Keyword: MT7 3-3	☐ Homework Help Online Keyword: MT7 3-3
☐ *Lesson Tutorial Video* 3-3	☐ *Lesson Tutorial Video* 3-3	☐ *Lesson Tutorial Video* 3-3	☐ *Lesson Tutorial Video* 3-3
☐ Reading Strategies 3-3 CRB	☐ Problem Solving 3-3 CRB	☐ Problem Solving 3-3 CRB	☐ Reading Strategies 3-3 CRB
☐ *Questioning Strategies* pp. 38–39			
☐ *IDEA Works!* 3-3			☐ *Multilingual Glossary*

ASSESSMENT
☐ Lesson Quiz, TE p. 131 and DT 3-3   ☐ State-Specific Test Prep Online Keyword: MT7 TestPrep

Teacher's Name _____ Class _____ Date _____

Lesson Plan 3-4
Functions pp. 134–137   Day _____

Objective Students represent functions with tables, graphs, or equations.

> **NCTM Standards:** Create and use representations to organize, record, and communicate mathematical ideas.

Pacing
☐ 45-minute Classes: 1 day   ☐ 90-minute Classes: 1/2 day   ☐ Other_____

WARM UP
☐ Warm Up TE p. 134 and Daily Transparency 3-4
☐ Problem of the Day TE p. 134 and Daily Transparency 3-4
☐ Countdown to Testing Transparency Week 5

TEACH
☐ Lesson Presentation CD-ROM 3-4
☐ Alternate Opener, Explorations Transparency 3-4, TE p. 134, and Exploration 3-4
☐ Reaching All Learners TE p. 135
☐ Teaching Transparency 3-4
☐ *Know-It Notebook* 3-4

PRACTICE AND APPLY
☐ Example 1: Average: 1–4, 8–11, 20, 25–35   Advanced: 8–11, 15–18, 22, 24–35
☐ Example 2: Average: 1–14, 16–19, 20, 25–35   Advanced: 8–16, 20–35

REACHING ALL LEARNERS – Differentiated Instruction for students with

Developing Knowledge	On-level Knowledge	Advanced Knowledge	English Language Development
☐ Inclusion TE p. 135	☐ Critical Thinking TE p. 135	☐ Critical Thinking TE p. 135	☐ Critical Thinking TE p. 135
☐ Practice A 3-4 CRB	☐ Practice B 3-4 CRB	☐ Practice C 3-4 CRB	☐ Practice A, B, or C 3-4 CRB
☐ Reteach 3-4 CRB	☐ Puzzles, Twisters & Teasers 3-4 CRB	☐ Challenge 3-4 CRB	☐ *Success for ELL* 3-4
☐ Homework Help Online Keyword: MT7 3-4	☐ Homework Help Online Keyword: MT7 3-4	☐ Homework Help Online Keyword: MT7 3-4	☐ Homework Help Online Keyword: MT7 3-4
☐ *Lesson Tutorial Video* 3-4	☐ *Lesson Tutorial Video* 3-4	☐ *Lesson Tutorial Video* 3-4	☐ *Lesson Tutorial Video* 3-4
☐ Reading Strategies 3-4 CRB	☐ Problem Solving 3-4 CRB	☐ Problem Solving 3-4 CRB	☐ Reading Strategies 3-4 CRB
☐ *Questioning Strategies* pp. 40–41			☐ Lesson Vocabulary SE p. 134
☐ *IDEA Works!* 3-4			☐ *Multilingual Glossary*

ASSESSMENT
☐ Lesson Quiz, TE p. 137 and DT 3-4   ☐ State-Specific Test Prep Online Keyword: MT7 TestPrep

Teacher's Name _____ Class _____ Date _____

Lesson Plan 3-5
Equations, Tables, and Graphs pp. 138–141   Day _____

Objective Students generate different representations of the same data.

NCTM Standards: Compute fluently and make reasonable estimates.

Pacing
☐ 45-minute Classes: 1 day   ☐ 90-minute Classes: 1/2 day   ☐ Other_____

WARM UP
☐ Warm Up TE p. 138 and Daily Transparency 3-5
☐ Problem of the Day TE p. 138 and Daily Transparency 3-5
☐ Countdown to Testing Transparency Week 5

TEACH
☐ Lesson Presentation CD-ROM 3-5
☐ Alternate Opener, Explorations Transparency 3-5, TE p. 138, and Exploration 3-5
☐ Reaching All Learners TE p. 139
☐ *Hands-On Lab Activities* 3-5
☐ *Know-It Notebook* 3-5

PRACTICE AND APPLY
☐ Example 1: Average: 1, 4, 7, 8, 12–20 Advanced: 4, 7, 8, 10–20
☐ Example 2: Average: 1, 2, 4, 5, 7, 8, 12–20 Advanced: 4, 5, 7–20
☐ Example 3: Average: 1–8, 12–20 Advanced: 4–20

REACHING ALL LEARNERS – Differentiated Instruction for students with

Developing Knowledge	On-level Knowledge	Advanced Knowledge	English Language Development
☐ Cooperative Learning TE p. 139	☐ Cooperative Learning TE p. 139	☐ Cooperative Learning TE p. 139	☐ Cooperative Learning TE p. 139
☐ Practice A 3-5 CRB	☐ Practice B 3-5 CRB	☐ Practice C 3-5 CRB	☐ Practice A, B, or C 3-5 CRB
☐ Reteach 3-5 CRB	☐ Puzzles, Twisters & Teasers 3-5 CRB	☐ Challenge 3-5 CRB	☐ *Success for ELL* 3-5
☐ Homework Help Online Keyword: MT7 3-5	☐ Homework Help Online Keyword: MT7 3-5	☐ Homework Help Online Keyword: MT7 3-5	☐ Homework Help Online Keyword: MT7 3-5
☐ *Lesson Tutorial Video* 3-5	☐ *Lesson Tutorial Video* 3-5	☐ *Lesson Tutorial Video* 3-5	☐ *Lesson Tutorial Video* 3-5
☐ Reading Strategies 3-5 CRB	☐ Problem Solving 3-5 CRB	☐ Problem Solving 3-5 CRB	☐ Reading Strategies 3-5 CRB
☐ Questioning Strategies pp. 42–43			
☐ *IDEA Works!* 3-5			☐ *Multilingual Glossary*

ASSESSMENT
☐ Lesson Quiz, TE p. 141 and DT 3-5   ☐ State-Specific Test Prep Online Keyword: MT7 TestPrep

Teacher's Name _____ Class _____ Date _____

Lesson Plan 3-6
Arithmetic Sequences pp. 142–145   Day _____

Objective Students identify and evaluate arithmetic sequences.

> **NCTM Standards:** Understand patterns, relations, and functions.

Pacing
- [ ] 45-minute Classes: 1 day
- [ ] 90-minute Classes: 1/2 day
- [ ] Other _____

WARM UP
- [ ] Warm Up TE p. 142 and Daily Transparency 3-6
- [ ] Problem of the Day TE p. 142 and Daily Transparency 3-6
- [ ] Countdown to Testing Transparency Week 5

TEACH
- [ ] Lesson Presentation CD-ROM 3-6
- [ ] Alternate Opener, Explorations Transparency 3-6, TE p. 142, and Exploration 3-6
- [ ] Reaching All Learners TE p. 143
- [ ] Teaching Transparency 3-6
- [ ] *Hands-On Lab Activities* 3-6
- [ ] *Know-It Notebook* 3-6

PRACTICE AND APPLY
- [ ] Example 1: Average: 1–6, 17–22, 48–55 Advanced: 17–22, 48–55
- [ ] Example 2: Average: 1–12, 17–28, 33–35, 48–55 Advanced: 17–28, 33–38, 45, 48–55
- [ ] Example 3: Average: 1–15, 17–31, 33–35, 41, 42, 48–55 Advanced: 17–31, 33–42, 45–55
- [ ] Example 4: Average: 1–35, 41–44, 48–55 Advanced: 17–32, 36–38, 41–55

REACHING ALL LEARNERS – Differentiated Instruction for students with

Developing Knowledge	On-level Knowledge	Advanced Knowledge	English Language Development
☐ Multiple Representations TE p. 143	☐ Multiple Representations TE p. 143	☐ Multiple Representations TE p. 143	☐ Multiple Representations TE p. 143
☐ Practice A 3-6 CRB	☐ Practice B 3-6 CRB	☐ Practice C 3-6 CRB	☐ Practice A, B, or C 3-6 CRB
☐ Reteach 3-6 CRB	☐ Puzzles, Twisters & Teasers 3-6 CRB	☐ Challenge 3-6 CRB	☐ *Success for ELL* 3-6
☐ Homework Help Online Keyword: MT7 3-6	☐ Homework Help Online Keyword: MT7 3-6	☐ Homework Help Online Keyword: MT7 3-6	☐ Homework Help Online Keyword: MT7 3-6
☐ *Lesson Tutorial Video* 3-6	☐ *Lesson Tutorial Video* 3-6	☐ *Lesson Tutorial Video* 3-6	☐ *Lesson Tutorial Video* 3-6
☐ Reading Strategies 3-6 CRB	☐ Problem Solving 3-6 CRB	☐ Problem Solving 3-6 CRB	☐ Reading Strategies 3-6 CRB
☐ *Questioning Strategies* pp. 44–45			☐ *Lesson Vocabulary* SE p. 142
☐ *IDEA Works!* 3-6			☐ *Multilingual Glossary*

ASSESSMENT
- [ ] Lesson Quiz, TE p. 145 and DT 3-6   ☐ State-Specific Test Prep Online Keyword: MT7 TestPrep

Teacher's Name _____ Class _____ Date _____

Lesson Plan 4-1
Exponents pp. 162–165   Day _____

Objective Students evaluate expressions with exponents.

> **NCTM Standards:** Understand numbers, ways of representing numbers, relationships among numbers, and number systems; Compute fluently and make reasonable estimates.

Pacing
☐ 45-minute Classes: 1 day   ☐ 90-minute Classes: 1/2 day   ☐ Other_____

WARM UP
☐ Warm Up TE p. 162 and Daily Transparency 4-1
☐ Problem of the Day TE p. 162 and Daily Transparency 4-1
☐ Countdown to Testing Transparency Week 6

TEACH
☐ Lesson Presentation CD-ROM 4-1
☐ Alternate Opener, Explorations Transparency 4-1, TE p. 162, and Exploration 4-1
☐ Reaching All Learners TE p. 163
☐ *Know-It Notebook* 4-1

PRACTICE AND APPLY
☐ Example 1: Average: 1–4, 15–20, 51–61   Advanced: 15–20, 31–34, 51–61
☐ Example 2: Average: 1–9, 15–25, 51–61   Advanced: 15–25, 31–38, 49, 51–61
☐ Example 3: Average: 1–13, 15–29, 39–44, 51–61   Advanced: 15–29, 31, 32, 37–44, 46–61
☐ Example 4: Average: 1–30, 39–45, 51–61   Advanced: 15–32, 37–61

REACHING ALL LEARNERS – Differentiated Instruction for students with

Developing Knowledge	On-level Knowledge	Advanced Knowledge	English Language Development
☐ Concrete Manipulatives TE p. 163	☐ Concrete Manipulatives TE p. 163	☐ Concrete Manipulatives TE p. 163	☐ Concrete Manipulatives TE p. 163
☐ Practice A 4-1 CRB	☐ Practice B 4-1 CRB	☐ Practice C 4-1 CRB	☐ Practice A, B, or C 4-1 CRB
☐ Reteach 4-1 CRB	☐ Puzzles, Twisters & Teasers 4-1 CRB	☐ Challenge 4-1 CRB	☐ *Success for ELL* 4-1
☐ Homework Help Online Keyword: MT7 4-1	☐ Homework Help Online Keyword: MT7 4-1	☐ Homework Help Online Keyword: MT7 4-1	☐ Homework Help Online Keyword: MT7 4-1
☐ *Lesson Tutorial Video* 4-1	☐ *Lesson Tutorial Video* 4-1	☐ *Lesson Tutorial Video* 4-1	☐ *Lesson Tutorial Video* 4-1
☐ Reading Strategies 4-1 CRB	☐ Problem Solving 4-1 CRB	☐ Problem Solving 4-1 CRB	☐ Reading Strategies 4-1 CRB
☐ *Questioning Strategies* pp. 46–47	☐ Cognitive Strategies TE p. 163	☐ Cognitive Strategies TE p. 163	☐ Lesson Vocabulary SE p. 162
☐ *IDEA Works!* 4-1			☐ *Multilingual Glossary*

ASSESSMENT
☐ Lesson Quiz, TE p. 165 and DT 4-1   ☐ State-Specific Test Prep Online Keyword: MT7 TestPrep

Teacher's Name _____ Class _____ Date _____

Lesson Plan 4-2
Look for a Pattern in Integer Exponents   pp. 166–169   Day _____

Objective Students evaluate expressions with negative exponents and evaluate the zero exponent.

> **NCTM Standards:** Compute fluently and make reasonable estimates; Understand patterns, relations, and functions; Make and investigate mathematical conjectures

Pacing
- [ ] 45-minute Classes: 1 day
- [ ] 90-minute Classes: 1/2 day
- [ ] Other _____

WARM UP
- [ ] Warm Up TE p. 166 and Daily Transparency 4-2
- [ ] Problem of the Day TE p. 166 and Daily Transparency 4-2
- [ ] Countdown to Testing Transparency Week 6

TEACH
- [ ] Lesson Presentation CD-ROM 4-2
- [ ] Alternate Opener, Explorations Transparency 4-2, TE p. 166, and Exploration 4-2
- [ ] Reaching All Learners TE p. 167
- [ ] Teaching Transparency 4-2
- [ ] *Know-It Notebook* 4-2

PRACTICE AND APPLY
- [ ] Example 1: Average: 1–4, 13–16, 44, 49–58 Advanced: 13–16, 44, 49–58
- [ ] Example 2: Average: 1–8, 13–20, 37–40, 44, 49–58 Advanced: 13–20, 37–42, 44, 49–58
- [ ] Example 3: Average: 1–30, 34–40, 44–47, 49–58 Advanced: 13–24, 31–34, 39–58

REACHING ALL LEARNERS – Differentiated Instruction for students with

Developing Knowledge	On-level Knowledge	Advanced Knowledge	English Language Development
☐ Inclusion TE p. 167	☐ Multiple Representations TE p. 167	☐ Multiple Representations TE p. 167	☐ Multiple Representations TE p. 167
☐ Practice A 4-2 CRB	☐ Practice B 4-2 CRB	☐ Practice C 4-2 CRB	☐ Practice A, B, or C 4-2 CRB
☐ Reteach 4-2 CRB	☐ Puzzles, Twisters & Teasers 4-2 CRB	☐ Challenge 4-2 CRB	☐ *Success for ELL* 4-2
☐ Homework Help Online Keyword: MT7 4-2	☐ Homework Help Online Keyword: MT7 4-2	☐ Homework Help Online Keyword: MT7 4-2	☐ Homework Help Online Keyword: MT7 4-2
☐ *Lesson Tutorial Video* 4-2	☐ *Lesson Tutorial Video* 4-2	☐ *Lesson Tutorial Video* 4-2	☐ *Lesson Tutorial Video* 4-2
☐ Reading Strategies 4-2 CRB	☐ Problem Solving 4-2 CRB	☐ Problem Solving 4-2 CRB	☐ Reading Strategies 4-2 CRB
☐ *Questioning Strategies* pp. 48–49			
☐ *IDEA Works!* 4-2			☐ *Multilingual Glossary*

ASSESSMENT
- [ ] Lesson Quiz, TE p. 169 and DT 4-2
- [ ] State-Specific Test Prep Online Keyword: MT7 TestPrep

Teacher's Name _____ Class _____ Date _____

Lesson Plan 4-3
Properties of Exponents pp. 170–173   Day _____

Objective Students apply the properties of exponents.

NCTM Standards: Compute fluently and make reasonable estimates.

Pacing
☐ 45-minute Classes: 1 day   ☐ 90-minute Classes: 1/2 day   ☐ Other_____

WARM UP
☐ Warm Up TE p. 170 and Daily Transparency 4-3
☐ Problem of the Day TE p. 170 and Daily Transparency 4-3
☐ Countdown to Testing Transparency Week 6

TEACH
☐ Lesson Presentation CD-ROM 4-3
☐ Alternate Opener, Explorations Transparency 4-3, TE p. 170, and Exploration 4-3
☐ Reaching All Learners TE p. 171
☐ Teaching Transparency 4-3
☐ *Hands-On Lab Activities* 4-3
☐ *Know-It Notebook* 4-3

PRACTICE AND APPLY
☐ Example 1: Average: 1–4, 13–16, 26, 29, 30, 32, 36, 49–60 Advanced: 13–16, 26, 29, 30, 32, 36, 41, 45, 49–60
☐ Example 2: Average: 1–8, 13–20, 25–30, 32–36, 38, 49–60 Advanced: 13–20, 35–41, 43, 45–60
☐ Example 3: Average: Average: 1–36, 38, 49–60 Advanced: 13–24, 35–60

REACHING ALL LEARNERS – Differentiated Instruction for students with

Developing Knowledge	On-level Knowledge	Advanced Knowledge	English Language Development
☐ Critical Thinking TE p. 171	☐ Critical Thinking TE p. 171	☐ Critical Thinking TE p. 171	☐ Critical Thinking TE p. 171
☐ Practice A 4-3 CRB	☐ Practice B 4-3 CRB	☐ Practice C 4-3 CRB	☐ Practice A, B, or C 4-3 CRB
☐ Reteach 4-3 CRB	☐ Puzzles, Twisters & Teasers 4-3 CRB	☐ Challenge 4-3 CRB	☐ *Success for ELL* 4-3
☐ Homework Help Online Keyword: MT7 4-3	☐ Homework Help Online Keyword: MT7 4-3	☐ Homework Help Online Keyword: MT7 4-3	☐ Homework Help Online Keyword: MT7 4-3
☐ *Lesson Tutorial Video* 4-3	☐ *Lesson Tutorial Video* 4-3	☐ *Lesson Tutorial Video* 4-3	☐ *Lesson Tutorial Video* 4-3
☐ Reading Strategies 4-3 CRB	☐ Problem Solving 4-3 CRB	☐ Problem Solving 4-3 CRB	☐ Reading Strategies 4-3 CRB
☐ *Questioning Strategies* pp. 50–51			
☐ *IDEA Works!* 4-3			☐ *Multilingual Glossary*

ASSESSMENT
☐ Lesson Quiz, TE p. 173 and DT 4-3   ☐ State-Specific Test Prep Online Keyword: MT7 TestPrep

Teacher's Name _____ Class _____ Date _____

Lesson Plan 4-4
Scientific Notation pp. 174–178   Day _____

Objective Students express large and small numbers in scientific notation and compare two numbers written in scientific notation.

> **NCTM Standards:** Understand numbers, ways of representing numbers, relationships among numbers, and number systems; Select, apply, and translate among mathematical representations to solve problems.

Pacing
☐ 45-minute Classes: 1 day   ☐ 90-minute Classes: 1/2 day   ☐ Other_____

WARM UP
☐ Warm Up TE p. 174 and Daily Transparency 4-4
☐ Problem of the Day TE p. 174 and Daily Transparency 4-4
☐ Countdown to Testing Transparency Week 6

TEACH
☐ Lesson Presentation CD-ROM 4-4
☐ Alternate Opener, Explorations Transparency 4-4, TE p. 174, and Exploration 4-4
☐ Reaching All Learners TE p. 175
☐ *Technology Lab Activities* 4-4
☐ *Know-It Notebook* 4-4

PRACTICE AND APPLY
☐ Example 1: Average: 1–4, 11–14, 21–26, 34, 53–60 Advanced: 11–14, 27–32, 34, 50, 53–60
☐ Example 2: Average: 1–8, 11–18, 21–26, 34, 36–42, 53–60 Advanced: 11–18, 27–32, 34, 36, 43–48, 50, 53–60
☐ Example 3: Average: 1–9, 11–19, 21–26, 33, 34, 36–42, 53–60 Advanced: 11–19, 29–42, 50, 53–60
☐ Example 4: Average: 1–26, 33, 34, 36–42, 49, 51, 53–60 Advanced: 11–20, 29–36, 43–60

REACHING ALL LEARNERS – Differentiated Instruction for students with

Developing Knowledge	On-level Knowledge	Advanced Knowledge	English Language Development
☐ Home Connection TE p. 175	☐ Home Connection TE p. 175	☐ Home Connection TE p. 175	☐ Home Connection TE p. 175
☐ Practice A 4-4 CRB	☐ Practice B 4-4 CRB	☐ Practice C 4-4 CRB	☐ Practice A, B, or C 4-4 CRB
☐ Reteach 4-4 CRB	☐ Puzzles, Twisters & Teasers 4-4 CRB	☐ Challenge 4-4 CRB	☐ *Success for ELL* 4-4
☐ Homework Help Online Keyword: MT7 4-4	☐ Homework Help Online Keyword: MT7 4-4	☐ Homework Help Online Keyword: MT7 4-4	☐ Homework Help Online Keyword: MT7 4-4
☐ *Lesson Tutorial Video* 4-4	☐ *Lesson Tutorial Video* 4-4	☐ *Lesson Tutorial Video* 4-4	☐ *Lesson Tutorial Video* 4-4
☐ Reading Strategies 4-4 CRB	☐ Problem Solving 4-4 CRB	☐ Problem Solving 4-4 CRB	☐ Reading Strategies 4-4 CRB
☐ *Questioning Strategies* pp. 52–53			☐ *Lesson Vocabulary* SE p. 174
☐ *IDEA Works!* 4-4			☐ *Multilingual Glossary*

ASSESSMENT
☐ Lesson Quiz, TE p. 178 and DT 4-4   ☐ State-Specific Test Prep Online Keyword: MT7 TestPrep

Teacher's Name _____ Class _____ Date _____

Lesson Plan 4-5
Squares and Square Roots pp. 182–185   Day _____

Objective Students find square roots.

> **NCTM Standards:** Understand meanings of operations and how they relate to one another.

Pacing
☐ 45-minute Classes: 1 day   ☐ 90-minute Classes: 1/2 day   ☐ Other_____

WARM UP
☐ Warm Up TE p. 182 and Daily Transparency 4-5
☐ Problem of the Day TE p. 182 and Daily Transparency 4-5
☐ Countdown to Testing Transparency Week 7

TEACH
☐ Lesson Presentation CD-ROM 4-5
☐ Alternate Opener, Explorations Transparency 4-5, TE p. 182, and Exploration 4-5
☐ Reaching All Learners TE p. 183
☐ *Know-It Notebook* 4-5

PRACTICE AND APPLY
☐ Example 1: Average: 1–8, 14–21, 27–30, 38–41, 52–61 Advanced: 14–21, 31–34, 40–45, 49, 52–61
☐ Example 2: Average: 1–9, 14–22, 27–30, 36–41, 47, 52–61 Advanced: 14–22, 31–49, 52–61
☐ Example 3: Average: 1–30, 36–41, 47, 50, 52–61 Advanced: 14–26, 31–37, 40–61

REACHING ALL LEARNERS – Differentiated Instruction for students with

Developing Knowledge	On-level Knowledge	Advanced Knowledge	English Language Development
☐ Cooperative Learning TE p. 183	☐ Cooperative Learning TE p. 183	☐ Cooperative Learning TE p. 183	☐ Cooperative Learning TE p. 183
☐ Practice A 4-5 CRB	☐ Practice B 4-5 CRB	☐ Practice C 4-5 CRB	☐ Practice A, B, or C 4-5 CRB
☐ Reteach 4-5 CRB	☐ Puzzles, Twisters & Teasers 4-5 CRB	☐ Challenge 4-5 CRB	☐ *Success for ELL* 4-5
☐ Homework Help Online Keyword: MT7 4-5	☐ Homework Help Online Keyword: MT7 4-5	☐ Homework Help Online Keyword: MT7 4-5	☐ Homework Help Online Keyword: MT7 4-5
☐ *Lesson Tutorial Video* 4-5	☐ *Lesson Tutorial Video* 4-5	☐ *Lesson Tutorial Video* 4-5	☐ *Lesson Tutorial Video* 4-5
☐ Reading Strategies 4-5 CRB	☐ Problem Solving 4-5 CRB	☐ Problem Solving 4-5 CRB	☐ Reading Strategies 4-5 CRB
☐ *Questioning Strategies* pp. 54–55			☐ Lesson Vocabulary SE p. 182
☐ *IDEA Works!* 4-5			☐ *Multilingual Glossary*

ASSESSMENT
☐ Lesson Quiz, TE p. 185 and DT 4-5   ☐ State-Specific Test Prep Online Keyword: MT7 TestPrep

Teacher's Name _____ Class _____ Date _____

Lesson Plan 4-6
Estimating Square Roots pp. 186–189   Day _____

Objective Students estimate square roots to a given number of decimal places and solve problems using square roots.

> **NCTM Standards:** Understand meanings of operations and how they relate to one another; Compute fluently and make reasonable estimates.

Pacing
☐ 45-minute Classes: 1 day   ☐ 90-minute Classes: 1/2 day   ☐ Other_____

WARM UP
☐ Warm Up TE p. 186 and Daily Transparency 4-6
☐ Problem of the Day TE p. 186 and Daily Transparency 4-6
☐ Countdown to Testing Transparency Week 7

TEACH
☐ Lesson Presentation CD-ROM 4-6
☐ Alternate Opener, Explorations Transparency 4-6, TE p. 186, and Exploration 4-6
☐ Reaching All Learners TE p. 187
☐ *Technology Lab Activities* 4-6
☐ *Know-It Notebook* 4-6

PRACTICE AND APPLY
☐ Example 1: Average: 1–5, 12–16, 23–28, 40–49   Advanced: 12–16, 23–29, 40–49
☐ Example 2: Average: 1–6, 12–17, 23–28, 36, 37, 40–49   Advanced: 12–17, 23–28, 35–37, 38–49
☐ Example 3: Average: 1–34, 36, 37; 40–49   Advanced: 12–49

REACHING ALL LEARNERS – Differentiated Instruction for students with

Developing Knowledge	On-level Knowledge	Advanced Knowledge	English Language Development
☐ Kinesthetic Experience TE p. 187	☐ Kinesthetic Experience TE p. 187	☐ Kinesthetic Experience TE p. 187	☐ Kinesthetic Experience TE p. 187
☐ Practice A 4-6 CRB	☐ Practice B 4-6 CRB	☐ Practice C 4-6 CRB	☐ Practice A, B, or C 4-6 CRB
☐ Reteach 4-6 CRB	☐ Puzzles, Twisters & Teasers 4-6 CRB	☐ Challenge 4-6 CRB	☐ *Success for ELL* 4-6
☐ Homework Help Online Keyword: MT7 4-6	☐ Homework Help Online Keyword: MT7 4-6	☐ Homework Help Online Keyword: MT7 4-6	☐ Homework Help Online Keyword: MT7 4-6
☐ *Lesson Tutorial Video* 4-6	☐ *Lesson Tutorial Video* 4-6	☐ *Lesson Tutorial Video* 4-6	☐ *Lesson Tutorial Video* 4-6
☐ Reading Strategies 4-6 CRB	☐ Problem Solving 4-6 CRB	☐ Problem Solving 4-6 CRB	☐ Reading Strategies 4-6 CRB
☐ *Questioning Strategies* pp. 56–57			
☐ *IDEA Works!* 4-6			☐ *Multilingual Glossary*

ASSESSMENT
☐ Lesson Quiz, TE p. 189 and DT 4-6   ☐ State-Specific Test Prep Online Keyword: MT7 TestPrep

Teacher's Name _____ Class _____ Date _____

Lesson Plan 4-7
The Real Numbers pp. 191–194   Day _____

Objective Students determine if a number is rational or irrational.

> **NCTM Standards:** Understand numbers, ways of representing numbers, relationships among numbers, and number systems.

Pacing
☐ 45-minute Classes: 1 day   ☐ 90-minute Classes: 1/2 day   ☐ Other_____

WARM UP
☐ Warm Up TE p. 191 and Daily Transparency 4-7
☐ Problem of the Day TE p. 191 and Daily Transparency 4-7
☐ Countdown to Testing Transparency Week 7

TEACH
☐ Lesson Presentation CD-ROM 4-7
☐ Alternate Opener, Explorations Transparency 4-7, TE p. 191, and Exploration 4-7
☐ Reaching All Learners TE p. 192
☐ Teaching Transparency 4-7
☐ *Hands-On Lab Activities* 4-7
☐ *Technology Lab Activities* 4-7
☐ *Know-It Notebook* 4-7

PRACTICE AND APPLY
☐ Example 1: Average: 1–4, 16–19, 34–38, 63–73 Advanced: 16–19, 31–38, 54–60, 63–73
☐ Example 2: Average: 1–12, 16–27, 34–38, 43, 63–73 Advanced: 16–27, 39–45, 54–73
☐ Example 3: Average: 1–30, 35–38, 43–47, 63–73 Advanced: 16–30, 39–45, 48–73

REACHING ALL LEARNERS – Differentiated Instruction for students with

Developing Knowledge	On-level Knowledge	Advanced Knowledge	English Language Development
☐ Graphic Organizers TE p. 192	☐ Graphic Organizers TE p. 192	☐ Graphic Organizers TE p. 192	☐ Graphic Organizers TE p. 192
☐ Practice A 4-7 CRB	☐ Practice B 4-7 CRB	☐ Practice C 4-7 CRB	☐ Practice A, B, or C 4-7 CRB
☐ Reteach 4-7 CRB	☐ Puzzles, Twisters & Teasers 4-7 CRB	☐ Challenge 4-7 CRB	☐ *Success for ELL* 4-7
☐ Homework Help Online Keyword: MT7 4-7	☐ Homework Help Online Keyword: MT7 4-7	☐ Homework Help Online Keyword: MT7 4-7	☐ Homework Help Online Keyword: MT7 4-7
☐ *Lesson Tutorial Video* 4-7	☐ *Lesson Tutorial Video* 4-7	☐ *Lesson Tutorial Video* 4-7	☐ *Lesson Tutorial Video* 4-7
☐ Reading Strategies 4-7 CRB	☐ Problem Solving 4-7 CRB	☐ Problem Solving 4-7 CRB	☐ Reading Strategies 4-7 CRB
☐ *Questioning Strategies* pp. 58–59			☐ *Lesson Vocabulary* SE p. 191
☐ *IDEA Works!* 4-7			☐ *Multilingual Glossary*

ASSESSMENT
☐ Lesson Quiz, TE p. 194 and DT 4-7   ☐ State-Specific Test Prep Online Keyword: MT7 TestPrep

Copyright © Holt, Rinehart and Winston.
All rights reserved.

Holt Mathematics

Teacher's Name _____ Class _____ Date _____

Lesson Plan 4-8
The Pythagorean Theorem pp. 196–199   Day _____

Objective Students use the Pythagorean Theorem to solve problems.

> **NCTM Standards:** Analyze characteristics and properties of two- and three-dimensional geometric shapes and develop mathematical arguments about geometric relationships.

Pacing
☐ 45-minute Classes: 1 day   ☐ 90-minute Classes: 1/2 day   ☐ Other _____

WARM UP
☐ Warm Up TE p. 196 and Daily Transparency 4-8
☐ Problem of the Day TE p. 196 and Daily Transparency 4-8
☐ Countdown to Testing Transparency Week 7

TEACH
☐ Lesson Presentation CD-ROM 4-8
☐ Alternate Opener, Explorations Transparency 4-8, TE p. 196, and Exploration 4-8
☐ Reaching All Learners TE p. 197
☐ Teaching Transparency 4-8
☐ *Hands-On Lab Activities* 4-8
☐ *Technology Lab Activities* 4-8
☐ *Know-It Notebook* 4-8

PRACTICE AND APPLY
☐ Example 1: Average: 1–4, 9–12, 17, 19, 23–25, 37–48 Advanced: 9–12, 17, 19, 23–30, 35, 37–48
☐ Example 2: Average: 1–7, 9–15, 17–25, 37–48 Advanced: 9–15, 20–30, 35–48
☐ Example 3: Average: 1–25, 31, 32, 37–48 Advanced: 9–16, 20–48

REACHING ALL LEARNERS – Differentiated Instruction for students with

Developing Knowledge	On-level Knowledge	Advanced Knowledge	English Language Development
☐ Modeling TE p. 197	☐ Modeling TE p. 197	☐ Modeling TE p. 197	☐ Modeling TE p. 197
☐ Practice A 4-8 CRB	☐ Practice B 4-8 CRB	☐ Practice C 4-8 CRB	☐ Practice A, B, or C 4-8 CRB
☐ Reteach 4-8 CRB	☐ Puzzles, Twisters & Teasers 4-8 CRB	☐ Challenge 4-8 CRB	☐ *Success for ELL* 4-8
☐ Homework Help Online Keyword: MT7 4-8	☐ Homework Help Online Keyword: MT7 4-8	☐ Homework Help Online Keyword: MT7 4-8	☐ Homework Help Online Keyword: MT7 4-8
☐ *Lesson Tutorial Video* 4-8	☐ *Lesson Tutorial Video* 4-8	☐ *Lesson Tutorial Video* 4-8	☐ *Lesson Tutorial Video* 4-8
☐ Reading Strategies 4-8 CRB	☐ Problem Solving 4-8 CRB	☐ Problem Solving 4-8 CRB	☐ Reading Strategies 4-8 CRB
☐ *Questioning Strategies* pp. 60–61			☐ Lesson Vocabulary SE p. 196
☐ *IDEA Works!* 4-8			☐ *Multilingual Glossary*

ASSESSMENT
☐ Lesson Quiz, TE p. 199 and DT 4-8   ☐ State-Specific Test Prep Online Keyword: MT7 TestPrep

Teacher's Name _____ Class _____ Date _____

Lesson Plan 5-1
Ratios and Proportions pp. 216–219   Day _____

Objective Students find equivalent ratios to create proportions.

> **NCTM Standards:** Understand numbers, ways of representing numbers, relationships among numbers, and number systems; Select, apply, and translate among mathematical representations to solve problems.

Pacing
☐ 45-minute Classes: 1 day   ☐ 90-minute Classes: 1/2 day   ☐ Other_____

WARM UP
☐ Warm Up TE p. 216 and Daily Transparency 5-1
☐ Problem of the Day TE p. 216 and Daily Transparency 5-1
☐ Countdown to Testing Transparency Week 8

TEACH
☐ Lesson Presentation CD-ROM 5-1
☐ Alternate Opener, Explorations Transparency 5-1, TE p. 216, and Exploration 5-1
☐ Reaching All Learners TE p. 217
☐ Teaching Transparency 5-1
☐ *Hands-On Lab Activities* 5-1
☐ *Technology Lab Activities* 5-1
☐ *Know-It Notebook* 5-1

PRACTICE AND APPLY
☐ Example 1: Average: 1–5, 11–15, 37–47 Advanced: 11–15, 37–47
☐ Example 2: Average: 1–9, 11–19, 24–29, 37–47 Advanced: 11–19, 23, 26–32, 34–47
☐ Example 3: Average: 1–22, 24–29, 37–47 Advanced: 11–23, 28–47

REACHING ALL LEARNERS – Differentiated Instruction for students with

Developing Knowledge	On-level Knowledge	Advanced Knowledge	English Language Development
☐ Inclusion TE p. 217	☐ Cooperative Learning TE p. 217	☐ Cooperative Learning TE p. 217	☐ Cooperative Learning TE p. 217
☐ Practice A 5-1 CRB	☐ Practice B 5-1 CRB	☐ Practice C 5-1 CRB	☐ Practice A, B, or C 5-1 CRB
☐ Reteach 5-1 CRB	☐ Puzzles, Twisters & Teasers 5-1 CRB	☐ Challenge 5-1 CRB	☐ *Success for ELL* 5-1
☐ Homework Help Online Keyword: MT7 5-1	☐ Homework Help Online Keyword: MT7 5-1	☐ Homework Help Online Keyword: MT7 5-1	☐ Homework Help Online Keyword: MT7 5-1
☐ *Lesson Tutorial Video* 5-1	☐ *Lesson Tutorial Video* 5-1	☐ *Lesson Tutorial Video* 5-1	☐ *Lesson Tutorial Video* 5-1
☐ Reading Strategies 5-1 CRB	☐ Problem Solving 5-1 CRB	☐ Problem Solving 5-1 CRB	☐ Reading Strategies 5-1 CRB
☐ *Questioning Strategies* pp. 62–63			☐ Lesson Vocabulary SE p. 216
☐ *IDEA Works!* 5-1			☐ *Multilingual Glossary*

ASSESSMENT
☐ Lesson Quiz, TE p. 219 and DT 5-1   ☐ State-Specific Test Prep Online Keyword: MT7 TestPrep

Teacher's Name _____ Class _____ Date _____

Lesson Plan 5-2

Ratios, Rates, and Unit Rates pp. 220–223   Day _____

Objective Students work with rates and ratios.

> **NCTM Standards:** Compute fluently and make reasonable estimates; Apply appropriate techniques, tools, and formulas to determine measurements; Select, apply, and translate among mathematical representations to solve problems.

Pacing
- [ ] 45-minute Classes: 1 day   □ 90-minute Classes: 1/2 day   □ Other_____

WARM UP
- [ ] Warm Up TE p. 220 and Daily Transparency 5-2
- [ ] Problem of the Day TE p. 220 and Daily Transparency 5-2
- [ ] Countdown to Testing Transparency Week 8

TEACH
- [ ] Lesson Presentation CD-ROM 5-2
- [ ] Alternate Opener, Explorations Transparency 5-2, TE p. 220, and Exploration 5-2
- [ ] Reaching All Learners TE p. 221
- [ ] *Technology Lab Activities* 5-2
- [ ] *Know-It Notebook* 5-2

PRACTICE AND APPLY
- [ ] Example 1: Average: 1, 8, 16–19, 28, 32–41 Advanced: 1, 8, 16–19, 28, 29, 31–41
- [ ] Example 2: Average: 1, 2, 8, 9, 16–19, 28, 32–41 Advanced: 8, 9, 16–19, 26, 28, 29, 31–41
- [ ] Example 3: Average: 1–6, 8–13, 16–23, 27, 28, 32–41 Advanced: 8–13, 16–23, 26–29, 31–41
- [ ] Example 4: Average: 1–25, 27, 28, 32–41 Advanced: 8–41

REACHING ALL LEARNERS – Differentiated Instruction for students with

Developing Knowledge	On-level Knowledge	Advanced Knowledge	English Language Development
□ Home Connection TE p. 221	□ Home Connection TE p. 221	□ Home Connection TE p. 221	□ Home Connection TE p. 221
□ Practice A 5-2 CRB	□ Practice B 5-2 CRB	□ Practice C 5-2 CRB	□ Practice A, B, or C 5-2 CRB
□ Reteach 5-2 CRB	□ Puzzles, Twisters & Teasers 5-2 CRB	□ Challenge 5-2 CRB	□ *Success for ELL* 5-2
□ Homework Help Online Keyword: MT7 5-2	□ Homework Help Online Keyword: MT7 5-2	□ Homework Help Online Keyword: MT7 5-2	□ Homework Help Online Keyword: MT7 5-2
□ *Lesson Tutorial Video* 5-2	□ *Lesson Tutorial Video* 5-2	□ *Lesson Tutorial Video* 5-2	□ *Lesson Tutorial Video* 5-2
□ Reading Strategies 5-2 CRB	□ Problem Solving 5-2 CRB	□ Problem Solving 5-2 CRB	□ Reading Strategies 5-2 CRB
□ *Questioning Strategies* pp. 64–65	□ Cognitive Strategies TE p. 221	□ Cognitive Strategies TE p. 221	□ Lesson Vocabulary SE p. 220
□ *IDEA Works!* 5-2			□ *Multilingual Glossary*

ASSESSMENT
- [ ] Lesson Quiz, TE p. 223 and DT 5-2   □ State-Specific Test Prep Online Keyword: MT7 TestPrep

Teacher's Name _____ Class _____ Date _____

Lesson Plan 5-3
Dimensional Analysis pp. 224–228   Day _____

Objective Students use one or more conversion factors to solve ratio problems.

> **NCTM Standards:** Compute fluently and make reasonable estimates Understand measurable attributes of objects and the units, systems, and processes of measurement; Select, apply, and translate among mathematical representations to solve problems.

Pacing
☐ 45-minute Classes: 1 day   ☐ 90-minute Classes: 1/2 day   ☐ Other_____

WARM UP
☐ Warm Up TE p. 224 and Daily Transparency 5-3
☐ Problem of the Day TE p. 224 and Daily Transparency 5-3
☐ Countdown to Testing Transparency Week 8

TEACH
☐ Lesson Presentation CD-ROM 5-3
☐ Alternate Opener, Explorations Transparency 5-3, TE p. 224, and Exploration 5-3
☐ Reaching All Learners TE p. 225
☐ *Hands-On Lab Activities* 5-3
☐ *Know-It Notebook* 5-3

PRACTICE AND APPLY
☐ Example 1: Average: 1–3, 7–9, 31–45 Advanced: 7–9, 31–45
☐ Example 2: Average: 1–4, 7–10, 13–18, 21, 31–45 Advanced: 7–10, 13–18, 26, 27, 29, 31–45
☐ Example 3: Average: 1–5, 7–11, 13–21, 25, 31–45 Advanced: 7–11, 13–21, 23–45
☐ Example 4: Average: 1–23, 25, 31–45 Advanced: 7–45

REACHING ALL LEARNERS – Differentiated Instruction for students with

Developing Knowledge	On-level Knowledge	Advanced Knowledge	English Language Development
☐ Curriculum Integration TE p. 225	☐ Curriculum Integration TE p. 225	☐ Curriculum Integration TE p. 225	☐ Curriculum Integration TE p. 225
☐ Practice A 5-3 CRB	☐ Practice B 5-3 CRB	☐ Practice C 5-3 CRB	☐ Practice A, B, or C 5-3 CRB
☐ Reteach 5-3 CRB	☐ Puzzles, Twisters & Teasers 5-3 CRB	☐ Challenge 5-3 CRB	☐ *Success for ELL* 5-3
☐ Homework Help Online Keyword: MT7 5-3	☐ Homework Help Online Keyword: MT7 5-3	☐ Homework Help Online Keyword: MT7 5-3	☐ Homework Help Online Keyword: MT7 5-3
☐ *Lesson Tutorial Video* 5-3	☐ *Lesson Tutorial Video* 5-3	☐ *Lesson Tutorial Video* 5-3	☐ *Lesson Tutorial Video* 5-3
☐ Reading Strategies 5-3 CRB	☐ Problem Solving 5-3 CRB	☐ Problem Solving 5-3 CRB	☐ Reading Strategies 5-3 CRB
☐ *Questioning Strategies* pp. 66–67			☐ Lesson Vocabulary SE p. 224
☐ *IDEA Works!* 5-3			☐ *Multilingual Glossary*

ASSESSMENT
☐ Lesson Quiz, TE p. 228 and DT 5-3   ☐ State-Specific Test Prep Online Keyword: MT7 TestPrep

Teacher's Name _____ Class _____ Date _____

Lesson Plan 5-4
Solving Proportions pp. 229–233   Day _____

Objective Students solve proportions.

> **NCTM Standards:** Represent and analyze mathematical situations and structures using algebraic symbols; Analyze change in various contexts.

Pacing
☐ 45-minute Classes: 1 day   ☐ 90-minute Classes: 1/2 day   ☐ Other_____

WARM UP
☐ Warm Up TE p. 229 and Daily Transparency 5-4
☐ Problem of the Day TE p. 229 and Daily Transparency 5-4
☐ Countdown to Testing Transparency Week 8

TEACH
☐ Lesson Presentation CD-ROM 5-4
☐ Alternate Opener, Explorations Transparency 5-4, TE p. 229, and Exploration 5-4
☐ Reaching All Learners TE p. 230
☐ Teaching Transparency 5-4
☐ *Know-It Notebook* 5-4

PRACTICE AND APPLY
☐ Example 1: Average: 1–5, 18–22, 39–42, 53–58 Advanced: 18–22, 42–47, 51, 53–58
☐ Example 2: Average: 1–11, 18–28, 39–42, 53–58 Advanced: 18–28, 42–47, 53–58
☐ Example 3: Average: 1–15, 18–36, 39–42, 53–58 Advanced: 18–36, 42–47, 53–58
☐ Example 4: Average: 1–16, 18–37, 39–42, 53–58 Advanced: 18–37, 42–47, 49, 51, 53–58
☐ Example 5: Average: 1–42, 48, 50, 51, 53–58 Advanced: 18–38, 42–58

REACHING ALL LEARNERS – Differentiated Instruction for students with

Developing Knowledge	On-level Knowledge	Advanced Knowledge	English Language Development
☐ Diversity TE p. 230	☐ Diversity TE p. 230	☐ Diversity TE p. 230	☐ Diversity TE p. 230
☐ Practice A 5-4 CRB	☐ Practice B 5-4 CRB	☐ Practice C 5-4 CRB	☐ Practice A, B, or C 5-4 CRB
☐ Reteach 5-4 CRB	☐ Puzzles, Twisters & Teasers 5-4 CRB	☐ Challenge 5-4 CRB	☐ *Success for ELL* 5-4
☐ Homework Help Online Keyword: MT7 5-4	☐ Homework Help Online Keyword: MT7 5-4	☐ Homework Help Online Keyword: MT7 5-4	☐ Homework Help Online Keyword: MT7 5-4
☐ *Lesson Tutorial Video* 5-4	☐ *Lesson Tutorial Video* 5-4	☐ *Lesson Tutorial Video* 5-4	☐ *Lesson Tutorial Video* 5-4
☐ Reading Strategies 5-4 CRB	☐ Problem Solving 5-4 CRB	☐ Problem Solving 5-4 CRB	☐ Reading Strategies 5-4 CRB
☐ *Questioning Strategies* pp. 68–69			☐ Lesson Vocabulary SE p. 229
☐ *IDEA Works!* 5-4			☐ *Multilingual Glossary*

ASSESSMENT
☐ Lesson Quiz, TE p. 233 and DT 5-4   ☐ State-Specific Test Prep Online Keyword: MT7 TestPrep

Teacher's Name _____ Class _____ Date _____

Lesson Plan 5-5
Similar Figures pp. 238–241 Day _____

Objective Students determine whether figures are similar, use scale factors, and find missing dimensions in similar figures.

> **NCTM Standards:** Apply appropriate techniques, tools, and formulas to determine measurements.

Pacing
- [ ] 45-minute Classes: 1 day   [ ] 90-minute Classes: 1/2 day   [ ] Other_____

WARM UP
- [ ] Warm Up TE p. 238 and Daily Transparency 5-5
- [ ] Problem of the Day TE p. 238 and Daily Transparency 5-5
- [ ] Countdown to Testing Transparency Week 9

TEACH
- [ ] Lesson Presentation CD-ROM 5-5
- [ ] Alternate Opener, Explorations Transparency 5-5, TE p. 238, and Exploration 5-5
- [ ] Reaching All Learners TE p. 239
- [ ] Teaching Transparency 5-5
- [ ] *Hands-On Lab Activities* 5-5
- [ ] *Know-It Notebook* 5-5

PRACTICE AND APPLY
- [ ] Example 1: Average: 1, 4, 7–9, 21–28 Advanced: 4, 7–10, 19, 21–28
- [ ] Example 2: Average: 1, 2, 4, 5, 7–9, 12–16, 21–28 Advanced: 4, 5, 7–16, 18, 19, 21–28
- [ ] Example 3: Average: 1–9, 12–17, 21–28 Advanced: 4–28

REACHING ALL LEARNERS – Differentiated Instruction for students with

Developing Knowledge	On-level Knowledge	Advanced Knowledge	English Language Development
[ ] Critical Thinking TE p. 239	[ ] Critical Thinking TE p. 239	[ ] Critical Thinking TE p. 239	[ ] Critical Thinking TE p. 239
[ ] Practice A 5-5 CRB	[ ] Practice B 5-5 CRB	[ ] Practice C 5-5 CRB	[ ] Practice A, B, or C 5-5 CRB
[ ] Reteach 5-5 CRB	[ ] Puzzles, Twisters & Teasers 5-5 CRB	[ ] Challenge 5-5 CRB	[ ] *Success for ELL* 5-5
[ ] Homework Help Online Keyword: MT7 5-5	[ ] Homework Help Online Keyword: MT7 5-5	[ ] Homework Help Online Keyword: MT7 5-5	[ ] Homework Help Online Keyword: MT7 5-5
[ ] *Lesson Tutorial Video* 5-5	[ ] *Lesson Tutorial Video* 5-5	[ ] *Lesson Tutorial Video* 5-5	[ ] *Lesson Tutorial Video* 5-5
[ ] Reading Strategies 5-5 CRB	[ ] Problem Solving 5-5 CRB	[ ] Problem Solving 5-5 CRB	[ ] Reading Strategies 5-5 CRB
[ ] *Questioning Strategies* pp. 70–71	[ ] Visual TE p. 239	[ ] Visual TE p. 239	[ ] Lesson Vocabulary SE p. 238
[ ] *IDEA Works!* 5-5			[ ] *Multilingual Glossary*

ASSESSMENT
- [ ] Lesson Quiz, TE p. 241 and DT 5-5   [ ] State-Specific Test Prep Online Keyword: MT7 TestPrep

Teacher's Name _____ Class _____ Date _____

Lesson Plan 5-6
Dilations pp. 244–247   Day _____

Objective Students identify and create dilations of plane figures.

> **NCTM Standards:** Apply transformations and use symmetry to analyze mathematical situations; Use visualization, spatial reasoning, and geometric modeling to solve problems.

Pacing
- ☐ 45-minute Classes: 1 day   ☐ 90-minute Classes: 1/2 day   ☐ Other _____

WARM UP
- ☐ Warm Up TE p. 244 and Daily Transparency 5-6
- ☐ Problem of the Day TE p. 244 and Daily Transparency 5-6
- ☐ Countdown to Testing Transparency Week 9

TEACH
- ☐ Lesson Presentation CD-ROM 5-6
- ☐ Alternate Opener, Explorations Transparency 5-6, TE p. 244, and Exploration 5-6
- ☐ Reaching All Learners TE p. 245
- ☐ Teaching Transparency 5-6
- ☐ *Hands-On Lab Activities* 5-6
- ☐ *Technology Lab Activities* 5-6
- ☐ *Know-It Notebook* 5-6

PRACTICE AND APPLY
- ☐ Example 1: Average: 1, 2, 7, 8, 15, 18–23 Advanced: 7, 8, 15, 16, 18–23
- ☐ Example 2: Average: 1–4, 7–10, 15, 18–23 Advanced: 7–10, 14–16, 18–23
- ☐ Example 3: Average: 1–13, 15, 18–23 Advanced: 7–23

REACHING ALL LEARNERS – Differentiated Instruction for students with

Developing Knowledge	On-level Knowledge	Advanced Knowledge	English Language Development
☐ Multiple Representations TE p. 245	☐ Multiple Representations TE p. 245	☐ Multiple Representations TE p. 245	☐ Multiple Representations TE p. 245
☐ Practice A 5-6 CRB	☐ Practice B 5-6 CRB	☐ Practice C 5-6 CRB	☐ Practice A, B, or C 5-6 CRB
☐ Reteach 5-6 CRB	☐ Puzzles, Twisters & Teasers 5-6 CRB	☐ Challenge 5-6 CRB	☐ *Success for ELL* 5-6
☐ Homework Help Online Keyword: MT7 5-6	☐ Homework Help Online Keyword: MT7 5-6	☐ Homework Help Online Keyword: MT7 5-6	☐ Homework Help Online Keyword: MT7 5-6
☐ *Lesson Tutorial Video* 5-6	☐ *Lesson Tutorial Video* 5-6	☐ *Lesson Tutorial Video* 5-6	☐ *Lesson Tutorial Video* 5-6
☐ Reading Strategies 5-6 CRB	☐ Problem Solving 5-6 CRB	☐ Problem Solving 5-6 CRB	☐ Reading Strategies 5-6 CRB
☐ *Questioning Strategies* pp. 72–73			☐ *Lesson Vocabulary* SE p. 244
☐ *IDEA Works!* 5-6			☐ *Multilingual Glossary*

ASSESSMENT
- ☐ Lesson Quiz, TE p. 247 and DT 5-6   ☐ State-Specific Test Prep Online Keyword: MT7 TestPrep

Teacher's Name _____ Class _____ Date _____

Lesson Plan 5-7
Indirect Measurement pp. 248–251   Day _____

Objective Students find measures indirectly by applying the properties of similar figures.

> **NCTM Standards:** Understand meanings of operations and how they relate to one another; Select and use various types of reasoning and methods of proof.

Pacing
☐ 45-minute Classes: 1 day   ☐ 90-minute Classes: 1/2 day   ☐ Other_____

WARM UP
☐ Warm Up TE p. 248 and Daily Transparency 5-7
☐ Problem of the Day TE p. 248 and Daily Transparency 5-7
☐ Countdown to Testing Transparency Week 9

TEACH
☐ Lesson Presentation CD-ROM 5-7
☐ Alternate Opener, Explorations Transparency 5-7, TE p. 248, and Exploration 5-7
☐ Reaching All Learners TE p. 249
☐ *Know-It Notebook* 5-7

PRACTICE AND APPLY
☐ Example 1: Average: 1, 4, 7–9, 15–24 Advanced: 4, 7–9, 12, 13, 15–24
☐ Example 2: Average: 1–11, 15–24 Advanced: 4–24

REACHING ALL LEARNERS – Differentiated Instruction for students with

Developing Knowledge	On-level Knowledge	Advanced Knowledge	English Language Development
☐ Kinesthetic Experience TE p. 249	☐ Kinesthetic Experience TE p. 249	☐ Kinesthetic Experience TE p. 249	☐ Kinesthetic Experience TE p. 249
☐ Practice A 5-7 CRB	☐ Practice B 5-7 CRB	☐ Practice C 5-7 CRB	☐ Practice A, B, or C 5-7 CRB
☐ Reteach 5-7 CRB	☐ Puzzles, Twisters & Teasers 5-7 CRB	☐ Challenge 5-7 CRB	☐ *Success for ELL* 5-7
☐ Homework Help Online Keyword: MT7 5-7	☐ Homework Help Online Keyword: MT7 5-7	☐ Homework Help Online Keyword: MT7 5-7	☐ Homework Help Online Keyword: MT7 5-7
☐ *Lesson Tutorial Video* 5-7	☐ *Lesson Tutorial Video* 5-7	☐ *Lesson Tutorial Video* 5-7	☐ *Lesson Tutorial Video* 5-7
☐ Reading Strategies 5-7 CRB	☐ Problem Solving 5-7 CRB	☐ Problem Solving 5-7 CRB	☐ Reading Strategies 5-7 CRB
☐ *Questioning Strategies* pp. 74–75	☐ Visual TE p. 249	☐ Visual TE p. 249	☐ Lesson Vocabulary SE p. 248
☐ *IDEA Works!* 5-7			☐ *Multilingual Glossary*

ASSESSMENT
☐ Lesson Quiz, TE p. 251 and DT 5-7   ☐ State-Specific Test Prep Online Keyword: MT7 TestPrep

Teacher's Name _____ Class _____ Date _____

Lesson Plan 5-8
Scale Drawings and Scale Models pp. 252–255   Day _____

Objective Students make comparisons between and find dimensions of scale drawings, models, and actual objects.

> **NCTM Standards:** Use visualization, spatial reasoning, and geometric modeling to solve problems; Understand measurable attributes of objects and the units, systems, and processes of measurement.

Pacing
☐ 45-minute Classes: 1 day   ☐ 90-minute Classes: 1/2 day   ☐ Other_____

WARM UP
☐ Warm Up TE p. 252 and Daily Transparency 5-8
☐ Problem of the Day TE p. 252 and Daily Transparency 5-8
☐ Countdown to Testing Transparency Week 9

TEACH
☐ Lesson Presentation CD-ROM 5-8
☐ Alternate Opener, Explorations Transparency 5-8, TE p. 252, and Exploration 5-8
☐ Reaching All Learners TE p. 253
☐ Teaching Transparency 5-8
☐ *Know-It Notebook* 5-8

PRACTICE AND APPLY
☐ Example 1: Average: 1, 5, 9–14, 31–40 Advanced: 5, 9–14, 31–40
☐ Example 2: Average: 1, 2, 5, 6, 9–14, 21, 31–40 Advanced: 5, 6, 9–14, 21, 22, 31–40
☐ Example 3: Average: 1–3, 5–7, 9–21, 24–29, 31–40 Advanced: 5–7, 12–17, 21–40
☐ Example 4: Average: 1–21, 24–29, 31–40 Advanced: 5–8, 12–17, 21–40

REACHING ALL LEARNERS – Differentiated Instruction for students with

Developing Knowledge	On-level Knowledge	Advanced Knowledge	English Language Development
☐ Modeling TE p. 253	☐ Modeling TE p. 253	☐ Modeling TE p. 253	☐ Modeling TE p. 253
☐ Practice A 5-8 CRB	☐ Practice B 5-8 CRB	☐ Practice C 5-8 CRB	☐ Practice A, B, or C 5-8 CRB
☐ Reteach 5-8 CRB	☐ Puzzles, Twisters & Teasers 5-8 CRB	☐ Challenge 5-8 CRB	☐ *Success for ELL* 5-8
☐ Homework Help Online Keyword: MT7 5-8	☐ Homework Help Online Keyword: MT7 5-8	☐ Homework Help Online Keyword: MT7 5-8	☐ Homework Help Online Keyword: MT7 5-8
☐ *Lesson Tutorial Video* 5-8	☐ *Lesson Tutorial Video* 5-8	☐ *Lesson Tutorial Video* 5-8	☐ *Lesson Tutorial Video* 5-8
☐ Reading Strategies 5-8 CRB	☐ Problem Solving 5-8 CRB	☐ Problem Solving 5-8 CRB	☐ Reading Strategies 5-8 CRB
☐ *Questioning Strategies* pp. 76–77			☐ Lesson Vocabulary SE p. 252
☐ *IDEA Works!* 5-8			☐ *Multilingual Glossary*

ASSESSMENT
☐ Lesson Quiz, TE p. 255 and DT 5-8   ☐ State-Specific Test Prep Online Keyword: MT7 TestPrep

Holt Mathematics

Teacher's Name _____ Class _____ Date _____

Lesson Plan 6-1
Relating Decimals, Fractions, and Percents pp. 274–277   Day _____

Objective Students compare and order decimals, fractions, and percents.

> **NCTM Standards:** Understand numbers, ways of representing numbers, relationships among numbers, and number systems; Select, apply, and translate among mathematical representations to solve problems.

Pacing
☐ 45-minute Classes: 1 day   ☐ 90-minute Classes: 1/2 day   ☐ Other_____

WARM UP
☐ Warm Up TE p. 274 and Daily Transparency 6-1
☐ Problem of the Day TE p. 274 and Daily Transparency 6-1
☐ Countdown to Testing Transparency Week 10

TEACH
☐ Lesson Presentation CD-ROM 6-1
☐ Alternate Opener, Explorations Transparency 6-1, TE p. 274, and Exploration 6-1
☐ Reaching All Learners TE p. 275
☐ Teaching Transparency 6-1
☐ *Hands-On Lab Activities* 6-1
☐ *Know-It Notebook* 6-1

PRACTICE AND APPLY
☐ Example 1: Average: 1–4, 12–15, 23–25, 32–40 Advanced: 12–15, 23–25, 27, 31–40
☐ Example 2: Average: 1–8, 12–19, 23–25, 32–40 Advanced: 12–19, 23–25, 27, 31–40
☐ Example 3: Average: 1–10, 12–21, 23–25, 30, 32–40 Advanced: 12–21, 23–27, 30–40
☐ Example 4: Average: 1–25, 28, 30, 32–40 Advanced: 12–40

REACHING ALL LEARNERS – Differentiated Instruction for students with

Developing Knowledge	On-level Knowledge	Advanced Knowledge	English Language Development
☐ Kinesthetic Experience TE p. 275	☐ Kinesthetic Experience TE p. 275	☐ Kinesthetic Experience TE p. 275	☐ Kinesthetic Experience TE p. 275
☐ Practice A 6-1 CRB	☐ Practice B 6-1 CRB	☐ Practice C 6-1 CRB	☐ Practice A, B, or C 6-1 CRB
☐ Reteach 6-1 CRB	☐ Puzzles, Twisters & Teasers 6-1 CRB	☐ Challenge 6-1 CRB	☐ *Success for ELL* 6-1
☐ Homework Help Online Keyword: MT7 6-1	☐ Homework Help Online Keyword: MT7 6-1	☐ Homework Help Online Keyword: MT7 6-1	☐ Homework Help Online Keyword: MT7 6-1
☐ *Lesson Tutorial Video* 6-1	☐ *Lesson Tutorial Video* 6-1	☐ *Lesson Tutorial Video* 6-1	☐ *Lesson Tutorial Video* 6-1
☐ Reading Strategies 6-1 CRB	☐ Problem Solving 6-1 CRB	☐ Problem Solving 6-1 CRB	☐ Reading Strategies 6-1 CRB
☐ *Questioning Strategies* pp. 78–79	☐ Reading Math TE p. 275	☐ Reading Math TE p. 275	☐ Lesson Vocabulary SE p. 274
☐ *IDEA Works!* 6-1			☐ *Multilingual Glossary*

ASSESSMENT
☐ Lesson Quiz, TE p. 277 and DT 6-1   ☐ State-Specific Test Prep Online Keyword: MT7 TestPrep

Holt Mathematics

Teacher's Name _____ Class _____ Date _____

Lesson Plan 6-2
Estimating with Percents pp. 278–282   Day _____

Objective Students estimate with percents.

> **NCTM Standards:** Understand numbers, ways of representing numbers, relationships among numbers, and number systems; Compute fluently and make reasonable estimates; Select, apply, and translate among mathematical representations to solve problems.

Pacing
☐ 45-minute Classes: 1 day   ☐ 90-minute Classes: 1/2 day   ☐ Other_____

WARM UP
☐ Warm Up TE p. 278 and Daily Transparency 6-2
☐ Problem of the Day TE p. 278 and Daily Transparency 6-2
☐ Countdown to Testing Transparency Week 10

TEACH
☐ Lesson Presentation CD-ROM 6-2
☐ Alternate Opener, Explorations Transparency 6-2, TE p. 278, and Exploration 6-2
☐ Reaching All Learners TE p. 279
☐ Teaching Transparency 6-2
☐ *Know-It Notebook* 6-2

PRACTICE AND APPLY
☐ Example 1: Average: 1–8, 11–18, 21–30, 47–63 Advanced: 11–18, 31–36, 45–63
☐ Example 2: Average: 1–9, 11–19, 21–30, 37–42, 47–63 Advanced: 11–19, 31–42, 44–63
☐ Example 3: Average: 1–30, 37–43, 47–63 Advanced: 11–20, 31–63

REACHING ALL LEARNERS – Differentiated Instruction for students with

Developing Knowledge	On-level Knowledge	Advanced Knowledge	English Language Development
☐ Inclusion TE p. 279	☐ Number Sense TE p. 279	☐ Number Sense TE p. 279	☐ Number Sense TE p. 279
☐ Practice A 6-2 CRB	☐ Practice B 6-2 CRB	☐ Practice C 6-2 CRB	☐ Practice A, B, or C 6-2 CRB
☐ Reteach 6-2 CRB	☐ Puzzles, Twisters & Teasers 6-2 CRB	☐ Challenge 6-2 CRB	☐ *Success for ELL* 6-2
☐ Homework Help Online Keyword: MT7 6-2	☐ Homework Help Online Keyword: MT7 6-2	☐ Homework Help Online Keyword: MT7 6-2	☐ Homework Help Online Keyword: MT7 6-2
☐ *Lesson Tutorial Video* 6-2	☐ *Lesson Tutorial Video* 6-2	☐ *Lesson Tutorial Video* 6-2	☐ *Lesson Tutorial Video* 6-2
☐ Reading Strategies 6-2 CRB	☐ Problem Solving 6-2 CRB	☐ Problem Solving 6-2 CRB	☐ Reading Strategies 6-2 CRB
☐ *Questioning Strategies* pp. 80–81			☐ Lesson Vocabulary SE p. 278
☐ *IDEA Works!* 6-2			☐ *Multilingual Glossary*

ASSESSMENT
☐ Lesson Quiz, TE p. 282 and DT 6-2   ☐ State-Specific Test Prep Online Keyword: MT7 TestPrep

Teacher's Name _____ Class _____ Date _____

Lesson Plan 6-3
Finding Percents pp. 283–287 Day _____

Objective Students find percents.

> **NCTM Standards:** Compute fluently and make reasonable estimates; Recognize and apply mathematics in contexts outside of mathematics; Select, apply, and translate among mathematical representations to solve problems.

Pacing
☐ 45-minute Classes: 1 day ☐ 90-minute Classes: 1/2 day ☐ Other_____

WARM UP
☐ Warm Up TE p. 283 and Daily Transparency 6-3
☐ Problem of the Day TE p. 283 and Daily Transparency 6-3
☐ Countdown to Testing Transparency Week 10

TEACH
☐ Lesson Presentation CD-ROM 6-3
☐ Alternate Opener, Explorations Transparency 6-3, TE p. 283, and Exploration 6-3
☐ Reaching All Learners TE p. 284
☐ *Hands-On Lab Activities* 6-3
☐ *Technology Lab Activities* 6-3
☐ *Know-It Notebook* 6-3

PRACTICE AND APPLY
☐ Example 1: Average: 1–4, 9–12, 29, 32, 35–47 Advanced: 9–12, 22–24, 29, 30, 32, 34–47
☐ Example 2: Average: 1–5, 9–13, 29, 32, 35–47 Advanced: 9–13, 22–24, 29–32, 34–47
☐ Example 3: Average: 1–21, 31, 35–47 Advanced: 9–15, 20–47
☐ Example 4: Average: 1–21, 31, 35–47 Advanced: 9–15, 20–47

REACHING ALL LEARNERS – Differentiated Instruction for students with

Developing Knowledge	On-level Knowledge	Advanced Knowledge	English Language Development
☐ Curriculum Integration TE p. 284	☐ Curriculum Integration TE p. 284	☐ Curriculum Integration TE p. 284	☐ Curriculum Integration TE p. 284
☐ Practice A 6-3 CRB	☐ Practice B 6-3 CRB	☐ Practice C 6-3 CRB	☐ Practice A, B, or C 6-3 CRB
☐ Reteach 6-3 CRB	☐ Puzzles, Twisters & Teasers 6-3 CRB	☐ Challenge 6-3 CRB	☐ *Success for ELL* 6-3
☐ Homework Help Online Keyword: MT7 6-3	☐ Homework Help Online Keyword: MT7 6-3	☐ Homework Help Online Keyword: MT7 6-3	☐ Homework Help Online Keyword: MT7 6-3
☐ *Lesson Tutorial Video* 6-3	☐ *Lesson Tutorial Video* 6-3	☐ *Lesson Tutorial Video* 6-3	☐ *Lesson Tutorial Video* 6-3
☐ Reading Strategies 6-3 CRB	☐ Problem Solving 6-3 CRB	☐ Problem Solving 6-3 CRB	☐ Reading Strategies 6-3 CRB
☐ *Questioning Strategies* pp. 82–83	☐ Reading Math TE p. 284	☐ Reading Math TE p. 284	
☐ *IDEA Works!* 6-3			☐ *Multilingual Glossary*

ASSESSMENT
☐ Lesson Quiz, TE p. 287 and DT 6-3 ☐ State-Specific Test Prep Online Keyword: MT7 TestPrep

Teacher's Name _____ Class _____ Date _____

Lesson Plan 6-4
Finding a Number When the Percent is Known pp. 288–291   Day _____

Objective Students find a number when the percent is known.

> **NCTM Standards:** Compute fluently and make reasonable estimates; Select, apply, and translate among mathematical representations to solve problems.

Pacing
☐ 45-minute Classes: 1 day   ☐ 90-minute Classes: 1/2 day   ☐ Other_____

WARM UP
☐ Warm Up TE p. 288 and Daily Transparency 6-4
☐ Problem of the Day TE p. 288 and Daily Transparency 6-4
☐ Countdown to Testing Transparency Week 10

TEACH
☐ Lesson Presentation CD-ROM 6-4
☐ Alternate Opener, Explorations Transparency 6-4, TE p. 288, and Exploration 6-4
☐ Reaching All Learners TE p. 289
☐ *Know-It Notebook* 6-4

PRACTICE AND APPLY
☐ Example 1: Average: 1–4, 7–10, 23–44 Advanced: 7–10, 13–15, 23–44
☐ Example 2: Average: 1–5, 7–11, 17–21, 23–44 Advanced: 7–11, 13–44
☐ Example 3: Average: 1–12, 17–21, 23–44 Advanced: 7–44

REACHING ALL LEARNERS – Differentiated Instruction for students with

Developing Knowledge	On-level Knowledge	Advanced Knowledge	English Language Development
☐ Critical Thinking TE p. 289	☐ Critical Thinking TE p. 289	☐ Critical Thinking TE p. 289	☐ Critical Thinking TE p. 289
☐ Practice A 6-4 CRB	☐ Practice B 6-4 CRB	☐ Practice C 6-4 CRB	☐ Practice A, B, or C 6-4 CRB
☐ Reteach 6-4 CRB	☐ Puzzles, Twisters & Teasers 6-4 CRB	☐ Challenge 6-4 CRB	☐ *Success for ELL* 6-4
☐ Homework Help Online Keyword: MT7 6-4	☐ Homework Help Online Keyword: MT7 6-4	☐ Homework Help Online Keyword: MT7 6-4	☐ Homework Help Online Keyword: MT7 6-4
☐ *Lesson Tutorial Video* 6-4	☐ *Lesson Tutorial Video* 6-4	☐ *Lesson Tutorial Video* 6-4	☐ *Lesson Tutorial Video* 6-4
☐ Reading Strategies 6-4 CRB	☐ Problem Solving 6-4 CRB	☐ Problem Solving 6-4 CRB	☐ Reading Strategies 6-4 CRB
☐ *Questioning Strategies* pp. 84–85	☐ Multiple Representations TE p. 289	☐ Multiple Representations TE p. 289	
☐ *IDEA Works!* 6-4			☐ *Multilingual Glossary*

ASSESSMENT
☐ Lesson Quiz, TE p. 291 and DT 6-4   ☐ State-Specific Test Prep Online Keyword: MT7 TestPrep

Teacher's Name _____ Class _____ Date _____

Lesson Plan 6-5
Percent Increase and Decrease pp. 294–297   Day _____

Objective Students find percent increase and decrease.

> **NCTM Standards:** Compute fluently and make reasonable estimates; Analyze change in various contexts; Select, apply, and translate among mathematical representations to solve problems.

Pacing
☐ 45-minute Classes: 1 day   ☐ 90-minute Classes: 1/2 day   ☐ Other_____

WARM UP
☐ Warm Up TE p. 294 and Daily Transparency 6-5
☐ Problem of the Day TE p. 294 and Daily Transparency 6-5
☐ Countdown to Testing Transparency Week 11

TEACH
☐ Lesson Presentation CD-ROM 6-5
☐ Alternate Opener, Explorations Transparency 6-5, TE p. 294, and Exploration 6-5
☐ Reaching All Learners TE p. 295
☐ *Know-It Notebook* 6-5

PRACTICE AND APPLY
☐ Example 1: Average: 1–3, 6–8, 11–13, 21, 22, 32–37 Advanced: 6–8, 14–16, 21, 22, 28, 30, 32–37
☐ Example 2: Average: 1–4, 6–9, 11–13, 21, 22, 25, 27, 32–37 Advanced: 6–9, 14–16, 21, 22, 27, 28, 30, 32–37
☐ Example 3: Average: 1–13, 17–22, 24, 25, 27, 32–37 Advanced: 6–37

REACHING ALL LEARNERS – Differentiated Instruction for students with

Developing Knowledge	On-level Knowledge	Advanced Knowledge	English Language Development
☐ Diversity TE p. 295	☐ Diversity TE p. 295	☐ Diversity TE p. 295	☐ Diversity TE p. 295
☐ Practice A 6-5 CRB	☐ Practice B 6-5 CRB	☐ Practice C 6-5 CRB	☐ Practice A, B, or C 6-5 CRB
☐ Reteach 6-5 CRB	☐ Puzzles, Twisters & Teasers 6-5 CRB	☐ Challenge 6-5 CRB	☐ *Success for ELL* 6-5
☐ Homework Help Online Keyword: MT7 6-5	☐ Homework Help Online Keyword: MT7 6-5	☐ Homework Help Online Keyword: MT7 6-5	☐ Homework Help Online Keyword: MT7 6-5
☐ *Lesson Tutorial Video* 6-5	☐ *Lesson Tutorial Video* 6-5	☐ *Lesson Tutorial Video* 6-5	☐ *Lesson Tutorial Video* 6-5
☐ Reading Strategies 6-5 CRB	☐ Problem Solving 6-5 CRB	☐ Problem Solving 6-5 CRB	☐ Reading Strategies 6-5 CRB
☐ *Questioning Strategies* pp. 86–87			☐ Lesson Vocabulary SE p. 294
☐ *IDEA Works!* 6-5			☐ *Multilingual Glossary*

ASSESSMENT
☐ Lesson Quiz, TE p. 297 and DT 6-5   ☐ State-Specific Test Prep Online Keyword: MT7 TestPrep

Teacher's Name _____ Class _____ Date _____

Lesson Plan 6-6
Applications of Percents pp. 298–301   Day _____

Objective Students find commission, sales tax, and percent of earnings.

> **NCTM Standards:** Compute fluently and make reasonable estimates; Select, apply, and translate among mathematical representations to solve problems.

Pacing
☐ 45-minute Classes: 1 day   ☐ 90-minute Classes: 1/2 day   ☐ Other_____

WARM UP
☐ Warm Up TE p. 298 and Daily Transparency 6-6
☐ Problem of the Day TE p. 298 and Daily Transparency 6-6
☐ Countdown to Testing Transparency Week 11

TEACH
☐ Lesson Presentation CD-ROM 6-6
☐ Alternate Opener, Explorations Transparency 6-6, TE p. 298, and Exploration 6-6
☐ Reaching All Learners TE p. 299
☐ *Technology Lab Activities* 6-6
☐ *Know-It Notebook* 6-6

PRACTICE AND APPLY
☐ Example 1: Average: 1, 5, 19–26 Advanced: 5, 15, 19–26
☐ Example 2: Average: 1, 2, 5, 6, 9–11, 19–26 Advanced: 5, 6, 9–11, 15, 19–26
☐ Example 3: Average: 1–3, 5–7, 14, 16, 19–26 Advanced: 5–7, 9–11, 14–26
☐ Example 4: Average: 1–14, 16, 19–26 Advanced: 5–26

REACHING ALL LEARNERS – Differentiated Instruction for students with

Developing Knowledge	On-level Knowledge	Advanced Knowledge	English Language Development
☐ Critical Thinking TE p. 299	☐ Critical Thinking TE p. 299	☐ Critical Thinking TE p. 299	☐ Critical Thinking TE p. 299
☐ Practice A 6-6 CRB	☐ Practice B 6-6 CRB	☐ Practice C 6-6 CRB	☐ Practice A, B, or C 6-6 CRB
☐ Reteach 6-6 CRB	☐ Puzzles, Twisters & Teasers 6-6 CRB	☐ Challenge 6-6 CRB	☐ *Success for ELL* 6-6
☐ Homework Help Online Keyword: MT7 6-6	☐ Homework Help Online Keyword: MT7 6-6	☐ Homework Help Online Keyword: MT7 6-6	☐ Homework Help Online Keyword: MT7 6-6
☐ *Lesson Tutorial Video* 6-6	☐ *Lesson Tutorial Video* 6-6	☐ *Lesson Tutorial Video* 6-6	☐ *Lesson Tutorial Video* 6-6
☐ Reading Strategies 6-6 CRB	☐ Problem Solving 6-6 CRB	☐ Problem Solving 6-6 CRB	☐ Reading Strategies 6-6 CRB
☐ *Questioning Strategies* pp. 88–89			☐ Lesson Vocabulary SE p. 298
☐ *IDEA Works!* 6-6			☐ *Multilingual Glossary*

ASSESSMENT
☐ Lesson Quiz, TE p. 301 and DT 6-6   ☐ State-Specific Test Prep Online Keyword: MT7 TestPrep

Holt Mathematics

Teacher's Name _____ Class _____ Date _____

Lesson Plan 6-7
Simple Interest pp. 302–305   Day _____

Objective Students compute simple interest.

> **NCTM Standards:** Compute fluently and make reasonable estimates; Select, apply, and translate among mathematical representations to solve problems.

Pacing
☐ 45-minute Classes: 1 day   ☐ 90-minute Classes: 1/2 day   ☐ Other_____

WARM UP
☐ Warm Up TE p. 302 and Daily Transparency 6-7
☐ Problem of the Day TE p. 302 and Daily Transparency 6-7
☐ Countdown to Testing Transparency Week 11

TEACH
☐ Lesson Presentation CD-ROM 6-7
☐ Alternate Opener, Explorations Transparency 6-7, TE p. 302, and Exploration 6-7
☐ Reaching All Learners TE p. 303
☐ Teaching Transparency 6-7
☐ *Know-It Notebook* 6-7

PRACTICE AND APPLY
☐ Example 1: Average: 1, 5, 9–12, 17, 22, 24–30 Advanced: 5, 13–17, 20, 22–30
☐ Example 2: Average: 1, 2, 5, 6, 9–12, 17, 18, 22, 24–30 Advanced: 5, 6, 13–18, 20, 22–30
☐ Example 3: Average: 1–3, 5–7, 9–12, 17, 18, 22, 24–30 Advanced: 5–7, 13–18, 20, 22–30
☐ Example 4: Average: 1–12, 17, 18, 22, 24–30 Advanced: 5–30

REACHING ALL LEARNERS – Differentiated Instruction for students with

Developing Knowledge	On-level Knowledge	Advanced Knowledge	English Language Development
☐ Cooperative Learning TE p. 303	☐ Cooperative Learning TE p. 303	☐ Cooperative Learning TE p. 303	☐ Cooperative Learning TE p. 303
☐ Practice A 6-7 CRB	☐ Practice B 6-7 CRB	☐ Practice C 6-7 CRB	☐ Practice A, B, or C 6-7 CRB
☐ Reteach 6-7 CRB	☐ Puzzles, Twisters & Teasers 6-7 CRB	☐ Challenge 6-7 CRB	☐ *Success for ELL* 6-7
☐ Homework Help Online Keyword: MT7 6-7	☐ Homework Help Online Keyword: MT7 6-7	☐ Homework Help Online Keyword: MT7 6-7	☐ Homework Help Online Keyword: MT7 6-7
☐ *Lesson Tutorial Video* 6-7	☐ *Lesson Tutorial Video* 6-7	☐ *Lesson Tutorial Video* 6-7	☐ *Lesson Tutorial Video* 6-7
☐ Reading Strategies 6-7 CRB	☐ Problem Solving 6-7 CRB	☐ Problem Solving 6-7 CRB	☐ Reading Strategies 6-7 CRB
☐ *Questioning Strategies* pp. 90–91			☐ Lesson Vocabulary SE p. 302
☐ *IDEA Works!* 6-7			☐ *Multilingual Glossary*

ASSESSMENT
☐ Lesson Quiz, TE p. 305 and DT 6-7   ☐ State-Specific Test Prep Online Keyword: MT7 TestPrep

Teacher's Name _____ Class _____ Date _____

Lesson Plan 7-1
Points, Lines, Planes, and Angles pp. 324–328   Day _____

Objective Students classify and name figures.

> **NCTM Standards:** Analyze characteristics and properties of two- and three-dimensional geometric shapes and develop mathematical arguments about geometric relationships.

Pacing
- [ ] 45-minute Classes: 1 day   [ ] 90-minute Classes: 1/2 day   [ ] Other_____

WARM UP
- [ ] Warm Up TE p. 324 and Daily Transparency 7-1
- [ ] Problem of the Day TE p. 324 and Daily Transparency 7-1
- [ ] Countdown to Testing Transparency Week 12

TEACH
- [ ] Lesson Presentation CD-ROM 7-1
- [ ] Alternate Opener, Explorations Transparency 7-1, TE p. 324, and Exploration 7-1
- [ ] Reaching All Learners TE p. 325
- [ ] Teaching Transparency 7-1
- [ ] *Hands-On Lab Activities* 7-1
- [ ] *Technology Lab Activities* 7-1
- [ ] *Know-It Notebook* 7-1

PRACTICE AND APPLY
- [ ] Example 1: Average: 1–5, 13–17, 25, 26, 40–46 Advanced: 13–17, 25, 26, 40–46
- [ ] Example 2: Average: 1–10, 13–22, 25–30, 40–46 Advanced: 13–22, 25–30, 40–46
- [ ] Example 3: Average: 1–34, 36, 37, 40–46 Advanced: 13–46

REACHING ALL LEARNERS – Differentiated Instruction for students with

Developing Knowledge	On-level Knowledge	Advanced Knowledge	English Language Development
[ ] Kinesthetic Experience TE p. 325	[ ] Kinesthetic Experience TE p. 325	[ ] Kinesthetic Experience TE p. 325	[ ] Kinesthetic Experience TE p. 325
[ ] Practice A 7-1 CRB	[ ] Practice B 7-1 CRB	[ ] Practice C 7-1 CRB	[ ] Practice A, B, or C 7-1 CRB
[ ] Reteach 7-1 CRB	[ ] Puzzles, Twisters & Teasers 7-1 CRB	[ ] Challenge 7-1 CRB	[ ] *Success for ELL* 7-1
[ ] Homework Help Online Keyword: MT7 7-1	[ ] Homework Help Online Keyword: MT7 7-1	[ ] Homework Help Online Keyword: MT7 7-1	[ ] Homework Help Online Keyword: MT7 7-1
[ ] *Lesson Tutorial Video* 7-1	[ ] *Lesson Tutorial Video* 7-1	[ ] *Lesson Tutorial Video* 7-1	[ ] *Lesson Tutorial Video* 7-1
[ ] Reading Strategies 7-1 CRB	[ ] Problem Solving 7-1 CRB	[ ] Problem Solving 7-1 CRB	[ ] Reading Strategies 7-1 CRB
[ ] *Questioning Strategies* pp. 92–93	[ ] Communicating Math TE p. 325	[ ] Communicating Math TE p. 325	[ ] Lesson Vocabulary SE p. 324
[ ] *IDEA Works!* 7-1			[ ] *Multilingual Glossary*

ASSESSMENT
- [ ] Lesson Quiz, TE p. 328 and DT 7-1   [ ] State-Specific Test Prep Online Keyword: MT7 TestPrep

Teacher's Name _____ Class _____ Date _____

Lesson Plan 7-2
Parallel and Perpendicular Lines pp. 330–333   Day _____

Objective Students identify parallel and perpendicular lines and the angles formed by a transversal.

> **NCTM Standards:** Analyze characteristics and properties of two- and three-dimensional geometric shapes and develop mathematical arguments about geometric relationships.

Pacing
☐ 45-minute Classes: 1 day   ☐ 90-minute Classes: 1/2 day   ☐ Other_____

WARM UP
☐ Warm Up TE p. 330 and Daily Transparency 7-2
☐ Problem of the Day TE p. 330 and Daily Transparency 7-2
☐ Countdown to Testing Transparency Week 12

TEACH
☐ Lesson Presentation CD-ROM 7-2
☐ Alternate Opener, Explorations Transparency 7-2, TE p. 330, and Exploration 7-2
☐ Reaching All Learners TE p. 331
☐ Teaching Transparency 7-2
☐ *Hands-On Lab Activities* 7-2
☐ *Know-It Notebook* 7-2

PRACTICE AND APPLY
☐ Example 1: Average: 1, 6, 26–33 Advanced: 6, 18–20, 26–33
☐ Example 2: Average: 1–18, 22, 26–33 Advanced: 6–33

REACHING ALL LEARNERS – Differentiated Instruction for students with

Developing Knowledge	On-level Knowledge	Advanced Knowledge	English Language Development
☐ Auditory Cues TE p. 331	☐ Auditory Cues TE p. 331	☐ Auditory Cues TE p. 331	☐ Auditory Cues TE p. 331
☐ Practice A 7-2 CRB	☐ Practice B 7-2 CRB	☐ Practice C 7-2 CRB	☐ Practice A, B, or C 7-2 CRB
☐ Reteach 7-2 CRB	☐ Puzzles, Twisters & Teasers 7-2 CRB	☐ Challenge 7-2 CRB	☐ *Success for ELL* 7-2
☐ Homework Help Online Keyword: MT7 7-2	☐ Homework Help Online Keyword: MT7 7-2	☐ Homework Help Online Keyword: MT7 7-2	☐ Homework Help Online Keyword: MT7 7-2
☐ *Lesson Tutorial Video* 7-2	☐ *Lesson Tutorial Video* 7-2	☐ *Lesson Tutorial Video* 7-2	☐ *Lesson Tutorial Video* 7-2
☐ Reading Strategies 7-2 CRB	☐ Problem Solving 7-2 CRB	☐ Problem Solving 7-2 CRB	☐ Reading Strategies 7-2 CRB
☐ *Questioning Strategies* pp. 94–95			☐ Lesson Vocabulary SE p. 330
☐ *IDEA Works!* 7-2			☐ *Multilingual Glossary*

ASSESSMENT
☐ Lesson Quiz, TE p. 333 and DT 7-2   ☐ State-Specific Test Prep Online Keyword: MT7 TestPrep

Teacher's Name _____ Class _____ Date _____

Lesson Plan 7-3
Angles in Triangles   pp. 336–340   Day _____

Objective Students find unknown angles in triangles.

> **NCTM Standards:** Analyze characteristics and properties of two- and three-dimensional geometric shapes and develop mathematical arguments about geometric relationships.

Pacing
☐ 45-minute Classes: 1 day   ☐ 90-minute Classes: 1/2 day   ☐ Other_____

WARM UP
☐ Warm Up TE p. 336 and Daily Transparency 7-3
☐ Problem of the Day TE p. 336 and Daily Transparency 7-3
☐ Countdown to Testing Transparency Week 13

TEACH
☐ Lesson Presentation CD-ROM 7-3
☐ Alternate Opener, Explorations Transparency 7-3, TE p. 336, and Exploration 7-3
☐ Reaching All Learners TE p. 337
☐ Teaching Transparency 7-3
☐ *Hands-On Lab Activities* 7-3
☐ *Know-It Notebook* 7-3

PRACTICE AND APPLY
☐ Example 1: Average: 1–3, 8–10, 15–17, 34–43 Advanced: 8–10, 15–17, 34–43
☐ Example 2: Average: 1–6, 8–13, 15–26, 30, 34–43 Advanced: 8–13, 15–26, 30–43
☐ Example 3: Average: 1–27, 30, 34–43 Advanced: 8–43

REACHING ALL LEARNERS – Differentiated Instruction for students with

Developing Knowledge	On-level Knowledge	Advanced Knowledge	English Language Development
☐ Inclusion TE p. 337	☐ Critical Thinking TE p. 337	☐ Critical Thinking TE p. 337	☐ Critical Thinking TE p. 337
☐ Practice A 7-3 CRB	☐ Practice B 7-3 CRB	☐ Practice C 7-3 CRB	☐ Practice A, B, or C 7-3 CRB
☐ Reteach 7-3 CRB	☐ Puzzles, Twisters & Teasers 7-3 CRB	☐ Challenge 7-3 CRB	☐ *Success for ELL* 7-3
☐ Homework Help Online Keyword: MT7 7-3	☐ Homework Help Online Keyword: MT7 7-3	☐ Homework Help Online Keyword: MT7 7-3	☐ Homework Help Online Keyword: MT7 7-3
☐ *Lesson Tutorial Video* 7-3	☐ *Lesson Tutorial Video* 7-3	☐ *Lesson Tutorial Video* 7-3	☐ *Lesson Tutorial Video* 7-3
☐ Reading Strategies 7-3 CRB	☐ Problem Solving 7-3 CRB	☐ Problem Solving 7-3 CRB	☐ Reading Strategies 7-3 CRB
☐ *Questioning Strategies* pp. 96–97			☐ Lesson Vocabulary SE p. 336
☐ *IDEA Works!* 7-3			☐ *Multilingual Glossary*

ASSESSMENT
☐ Lesson Quiz, TE p. 340 and DT 7-3   ☐ State-Specific Test Prep Online Keyword: MT7 TestPrep

Teacher's Name _____ Class _____ Date _____

Lesson Plan 7-4
Classifying Polygons pp. 341–345   Day _____

Objective Students classify and find angles in polygons.

> **NCTM Standards:** Analyze characteristics and properties of two- and three-dimensional geometric shapes and develop mathematical arguments about geometric relationships.

Pacing
- ☐ 45-minute Classes: 1 day   ☐ 90-minute Classes: 1/2 day   ☐ Other _____

WARM UP
- ☐ Warm Up TE p. 341 and Daily Transparency 7-4
- ☐ Problem of the Day TE p. 341 and Daily Transparency 7-4
- ☐ Countdown to Testing Transparency Week 13

TEACH
- ☐ Lesson Presentation CD-ROM 7-4
- ☐ Alternate Opener, Explorations Transparency 7-4, TE p. 341, and Exploration 7-4
- ☐ Reaching All Learners TE p. 342
- ☐ Teaching Transparency 7-4
- ☐ *Hands-On Lab Activities* 7-4
- ☐ *Technology Lab Activities* 7-4
- ☐ *Know-It Notebook* 7-4

PRACTICE AND APPLY
- ☐ Example 1: Average: 1–3, 10–12, 43–49 Advanced: 10–12, 43–49
- ☐ Example 2: Average: 1–6, 10–15, 19–32, 43–49 Advanced: 10–15, 21–33, 38, 41, 43–49
- ☐ Example 3: Average: 1–32, 36, 37, 43–49 Advanced: 10–20, 24–49

REACHING ALL LEARNERS – Differentiated Instruction for students with

Developing Knowledge	On-level Knowledge	Advanced Knowledge	English Language Development
☐ Diversity TE p. 342	☐ Diversity TE p. 342	☐ Diversity TE p. 342	☐ Diversity TE p. 342
☐ Practice A 7-4 CRB	☐ Practice B 7-4 CRB	☐ Practice C 7-4 CRB	☐ Practice A, B, or C 7-4 CRB
☐ Reteach 7-4 CRB	☐ Puzzles, Twisters & Teasers 7-4 CRB	☐ Challenge 7-4 CRB	☐ *Success for ELL* 7-4
☐ Homework Help Online Keyword: MT7 7-4	☐ Homework Help Online Keyword: MT7 7-4	☐ Homework Help Online Keyword: MT7 7-4	☐ Homework Help Online Keyword: MT7 7-4
☐ *Lesson Tutorial Video* 7-4	☐ *Lesson Tutorial Video* 7-4	☐ *Lesson Tutorial Video* 7-4	☐ *Lesson Tutorial Video* 7-4
☐ Reading Strategies 7-4 CRB	☐ Problem Solving 7-4 CRB	☐ Problem Solving 7-4 CRB	☐ Reading Strategies 7-4 CRB
☐ *Questioning Strategies* pp. 98–99			☐ Lesson Vocabulary SE p. 341
☐ *IDEA Works!* 7-4			☐ *Multilingual Glossary*

ASSESSMENT
- ☐ Lesson Quiz, TE p. 345 and DT 7-4   ☐ State-Specific Test Prep Online Keyword: MT7 TestPrep

Teacher's Name _____ Class _____ Date _____

Lesson Plan 7-5
Coordinate Geometry pp. 347–351   Day _____

Objective Students identify polygons in the coordinate plane.

> **NCTM Standards:** Specify locations and describe spatial relationships using coordinate geometry and other representational systems.

Pacing
☐ 45-minute Classes: 1 day   ☐ 90-minute Classes: 1/2 day   ☐ Other_____

WARM UP
☐ Warm Up TE p. 347 and Daily Transparency 7-5
☐ Problem of the Day TE p. 347 and Daily Transparency 7-5
☐ Countdown to Testing Transparency Week 13

TEACH
☐ Lesson Presentation CD-ROM 7-5
☐ Alternate Opener, Explorations Transparency 7-5, TE p. 347, and Exploration 7-5
☐ Reaching All Learners TE p. 348
☐ Teaching Transparency 7-5
☐ *Technology Lab Activities* 7-5
☐ *Know-It Notebook* 7-5

PRACTICE AND APPLY
☐ Example 1: Average: 1–4, 11–14, 21–24, 44–53 Advanced: 11–14, 21–24, 27–28, 42, 44–53
☐ Example 2: Average: 1–6, 11–16, 21–24, 26, 44–53 Advanced: 11–16, 21–28, 42, 44–53
☐ Example 3: Average: 1–8, 11–18, 21–24, 26, 31–36, 44–53 Advanced: 11–18, 21–28, 30–53
☐ Example 4: Average: 1–18, 21–24, 26, 31–36, 44–53 Advanced: 11–53

REACHING ALL LEARNERS – Differentiated Instruction for students with

Developing Knowledge	On-level Knowledge	Advanced Knowledge	English Language Development
☐ Cooperative Learning TE p. 348	☐ Cooperative Learning TE p. 348	☐ Cooperative Learning TE p. 348	☐ Cooperative Learning TE p. 348
☐ Practice A 7-5 CRB	☐ Practice B 7-5 CRB	☐ Practice C 7-5 CRB	☐ Practice A, B, or C 7-5 CRB
☐ Reteach 7-5 CRB	☐ Puzzles, Twisters & Teasers 7-5 CRB	☐ Challenge 7-5 CRB	☐ *Success for ELL* 7-5
☐ Homework Help Online Keyword: MT7 7-5	☐ Homework Help Online Keyword: MT7 7-5	☐ Homework Help Online Keyword: MT7 7-5	☐ Homework Help Online Keyword: MT7 7-5
☐ *Lesson Tutorial Video* 7-5	☐ *Lesson Tutorial Video* 7-5	☐ *Lesson Tutorial Video* 7-5	☐ *Lesson Tutorial Video* 7-5
☐ Reading Strategies 7-5 CRB	☐ Problem Solving 7-5 CRB	☐ Problem Solving 7-5 CRB	☐ Reading Strategies 7-5 CRB
☐ *Questioning Strategies* pp. 100–101			☐ Lesson Vocabulary SE p. 347
☐ *IDEA Works!* 7-5			☐ *Multilingual Glossary*

ASSESSMENT
☐ Lesson Quiz, TE p. 351 and DT 7-5   ☐ State-Specific Test Prep Online Keyword: MT7 TestPrep

Teacher's Name _____ Class _____ Date _____

Lesson Plan 7-6
Congruence pp. 354–357   Day _____

Objective Students use properties of congruent figures to solve problems.

> **NCTM Standards:** Use visualization, spatial reasoning, and geometric modeling to solve problems.

Pacing
☐ 45-minute Classes: 1 day   ☐ 90-minute Classes: 1/2 day   ☐ Other_____

WARM UP
☐ Warm Up TE p. 354 and Daily Transparency 7-6
☐ Problem of the Day TE p. 354 and Daily Transparency 7-6
☐ Countdown to Testing Transparency Week 13

TEACH
☐ Lesson Presentation CD-ROM 7-6
☐ Alternate Opener, Explorations Transparency 7-6, TE p. 354, and Exploration 7-6
☐ Reaching All Learners TE p. 355
☐ *Hands-On Lab Activities* 7-6
☐ *Know-It Notebook* 7-6

PRACTICE AND APPLY
☐ Example 1: Average: 1, 2, 6, 7, 19–29 Advanced: 6, 7, 16, 19–29
☐ Example 2: Average: 1–14, 19–29 Advanced: 6–29

REACHING ALL LEARNERS – Differentiated Instruction for students with

Developing Knowledge	On-level Knowledge	Advanced Knowledge	English Language Development
☐ Concrete Manipulatives TE p. 355	☐ Concrete Manipulatives TE p. 355	☐ Concrete Manipulatives TE p. 355	☐ Concrete Manipulatives TE p. 355
☐ Practice A 7-6 CRB	☐ Practice B 7-6 CRB	☐ Practice C 7-6 CRB	☐ Practice A, B, or C 7-6 CRB
☐ Reteach 7-6 CRB	☐ Puzzles, Twisters & Teasers 7-6 CRB	☐ Challenge 7-6 CRB	☐ *Success for ELL* 7-6
☐ Homework Help Online Keyword: MT7 7-6	☐ Homework Help Online Keyword: MT7 7-6	☐ Homework Help Online Keyword: MT7 7-6	☐ Homework Help Online Keyword: MT7 7-6
☐ *Lesson Tutorial Video* 7-6	☐ *Lesson Tutorial Video* 7-6	☐ *Lesson Tutorial Video* 7-6	☐ *Lesson Tutorial Video* 7-6
☐ Reading Strategies 7-6 CRB	☐ Problem Solving 7-6 CRB	☐ Problem Solving 7-6 CRB	☐ Reading Strategies 7-6 CRB
☐ *Questioning Strategies* pp. 102–103	☐ Communicating Math TE p. 355	☐ Communicating Math TE p. 355	☐ Lesson Vocabulary SE p. 354
☐ *IDEA Works!* 7-6			☐ *Multilingual Glossary*

ASSESSMENT
☐ Lesson Quiz, TE p. 357 and DT 7-6   ☐ State-Specific Test Prep Online Keyword: MT7 TestPrep

Teacher's Name _____ Class _____ Date _____

Lesson Plan 7-7
Transformations pp. 358–361   Day _____

Objective Students transform plane figures using translations, rotations, and reflections.

> **NCTM Standards:** Analyze change in various contexts; Apply transformations and use symmetry to analyze mathematical situations; Use visualization, spatial reasoning, and geometric modeling to solve problems.

Pacing
☐ 45-minute Classes: 1 day   ☐ 90-minute Classes: 1/2 day   ☐ Other_____

WARM UP
☐ Warm Up TE p. 358 and Daily Transparency 7-7
☐ Problem of the Day TE p. 358 and Daily Transparency 7-7
☐ Countdown to Testing Transparency Week 14

TEACH
☐ Lesson Presentation CD-ROM 7-7
☐ Alternate Opener, Explorations Transparency 7-7, TE p. 358, and Exploration 7-7
☐ Reaching All Learners TE p. 359
☐ Teaching Transparency 7-7
☐ *Know-It Notebook* 7-7

PRACTICE AND APPLY
☐ Example 1: Average: 1, 2, 11, 12, 36–43 Advanced: 11, 12, 34, 36–43
☐ Example 2: Average: 1–6, 11–16, 21–23, 36–43 Advanced: 11–16, 21–23, 33–43
☐ Example 3: Average: 1–26, 30–32, 36–43 Advanced: 11–43

REACHING ALL LEARNERS – Differentiated Instruction for students with

Developing Knowledge	On-level Knowledge	Advanced Knowledge	English Language Development
☐ Concrete Manipulatives TE p. 359	☐ Concrete Manipulatives TE p. 359	☐ Concrete Manipulatives TE p. 359	☐ Concrete Manipulatives TE p. 359
☐ Practice A 7-7 CRB	☐ Practice B 7-7 CRB	☐ Practice C 7-7 CRB	☐ Practice A, B, or C 7-7 CRB
☐ Reteach 7-7 CRB	☐ Puzzles, Twisters & Teasers 7-7 CRB	☐ Challenge 7-7 CRB	☐ *Success for ELL* 7-7
☐ Homework Help Online Keyword: MT7 7-7	☐ Homework Help Online Keyword: MT7 7-7	☐ Homework Help Online Keyword: MT7 7-7	☐ Homework Help Online Keyword: MT7 7-7
☐ *Lesson Tutorial Video* 7-7	☐ *Lesson Tutorial Video* 7-7	☐ *Lesson Tutorial Video* 7-7	☐ *Lesson Tutorial Video* 7-7
☐ Reading Strategies 7-7 CRB	☐ Problem Solving 7-7 CRB	☐ Problem Solving 7-7 CRB	☐ Reading Strategies 7-7 CRB
☐ *Questioning Strategies* pp. 104–105			☐ Lesson Vocabulary SE p. 358
☐ *IDEA Works!* 7-7			☐ *Multilingual Glossary*

ASSESSMENT
☐ Lesson Quiz, TE p. 361 and DT 7-7   ☐ State-Specific Test Prep Online Keyword: MT7 TestPrep

Teacher's Name _____ Class _____ Date _____

Lesson Plan 7-8
Symmetry pp. 364–367 Day _____

Objective Students identify symmetry in figures.

NCTM Standards: Apply transformations and use symmetry to analyze mathematical situations; Use visualization, spatial reasoning, and geometric modeling to solve problems;

Pacing
☐ 45-minute Classes: 1 day ☐ 90-minute Classes: 1/2 day ☐ Other_____

WARM UP
☐ Warm Up TE p. 364 and Daily Transparency 7-8
☐ Problem of the Day TE p. 364 and Daily Transparency 7-8
☐ Countdown to Testing Transparency Week 14

TEACH
☐ Lesson Presentation CD-ROM 7-8
☐ Alternate Opener, Explorations Transparency 7-8, TE p. 364, and Exploration 7-8
☐ Reaching All Learners TE p. 365
☐ Teaching Transparency 7-8
☐ *Technology Lab Activities* 7-8
☐ *Know-It Notebook* 7-8

PRACTICE AND APPLY
☐ Example 1: Average: 1–4, 8–10, 19–22, 27–35 Advanced: 8–10, 19–22, 27–35
☐ Example 2: Average: 1–16, 19–23, 27–35 Advanced: 8–35

REACHING ALL LEARNERS – Differentiated Instruction for students with

Developing Knowledge	On-level Knowledge	Advanced Knowledge	English Language Development
☐ Concrete Manipulatives TE p. 365	☐ Concrete Manipulatives TE p. 365	☐ Concrete Manipulatives TE p. 365	☐ Concrete Manipulatives TE p. 365
☐ Practice A 7-8 CRB	☐ Practice B 7-8 CRB	☐ Practice C 7-8 CRB	☐ Practice A, B, or C 7-8 CRB
☐ Reteach 7-8 CRB	☐ Puzzles, Twisters & Teasers 7-8 CRB	☐ Challenge 7-8 CRB	☐ *Success for ELL* 7-8
☐ Homework Help Online Keyword: MT7 7-8	☐ Homework Help Online Keyword: MT7 7-8	☐ Homework Help Online Keyword: MT7 7-8	☐ Homework Help Online Keyword: MT7 7-8
☐ *Lesson Tutorial Video* 7-8	☐ *Lesson Tutorial Video* 7-8	☐ *Lesson Tutorial Video* 7-8	☐ *Lesson Tutorial Video* 7-8
☐ Reading Strategies 7-8 CRB	☐ Problem Solving 7-8 CRB	☐ Problem Solving 7-8 CRB	☐ Reading Strategies 7-8 CRB
☐ *Questioning Strategies* pp. 106–107			☐ Lesson Vocabulary SE p. 364
☐ *IDEA Works!* 7-8			☐ *Multilingual Glossary*

ASSESSMENT
☐ Lesson Quiz, TE p. 367 and DT 7-8 ☐ State-Specific Test Prep Online Keyword: MT7 TestPrep

Teacher's Name _____ Class _____ Date _____

Lesson Plan 7-9
Tessellations pp. 368–371   Day _____

Objective Students create tessellations.

> **NCTM Standards:** Apply transformations and use symmetry to analyze mathematical situations; Use visualization, spatial reasoning, and geometric modeling to solve problems.

Pacing
☐ 45-minute Classes: 1 day   ☐ 90-minute Classes: 1/2 day   ☐ Other_____

WARM UP
☐ Warm Up TE p. 368 and Daily Transparency 7-9
☐ Problem of the Day TE p. 368 and Daily Transparency 7-9
☐ Countdown to Testing Transparency Week 14

TEACH
☐ Lesson Presentation CD-ROM 7-9
☐ Alternate Opener, Explorations Transparency 7-9, TE p. 368, and Exploration 7-9
☐ Reaching All Learners TE p. 369
☐ Teaching Transparency 7-9
☐ *Know-It Notebook* 7-9

PRACTICE AND APPLY
☐ Example 1: Average: 1, 3, 5–10, 17–23 Advanced: 3, 8–13, 17–23
☐ Example 2: Average: 1–10, 14, 17–23 Advanced: 3, 4, 8–23

REACHING ALL LEARNERS – Differentiated Instruction for students with

Developing Knowledge	On-level Knowledge	Advanced Knowledge	English Language Development
☐ Modeling TE p. 369	☐ Modeling TE p. 369	☐ Modeling TE p. 369	☐ Modeling TE p. 369
☐ Practice A 7-9 CRB	☐ Practice B 7-9 CRB	☐ Practice C 7-9 CRB	☐ Practice A, B, or C 7-9 CRB
☐ Reteach 7-9 CRB	☐ Puzzles, Twisters & Teasers 7-9 CRB	☐ Challenge 7-9 CRB	☐ *Success for ELL* 7-9
☐ Homework Help Online Keyword: MT7 7-9	☐ Homework Help Online Keyword: MT7 7-9	☐ Homework Help Online Keyword: MT7 7-9	☐ Homework Help Online Keyword: MT7 7-9
☐ *Lesson Tutorial Video* 7-9	☐ *Lesson Tutorial Video* 7-9	☐ *Lesson Tutorial Video* 7-9	☐ *Lesson Tutorial Video* 7-9
☐ Reading Strategies 7-9 CRB	☐ Problem Solving 7-9 CRB	☐ Problem Solving 7-9 CRB	☐ Reading Strategies 7-9 CRB
☐ *Questioning Strategies* pp. 108–109			☐ Lesson Vocabulary SE p. 368
☐ *IDEA Works!* 7-9			☐ *Multilingual Glossary*

ASSESSMENT
☐ Lesson Quiz, TE p. 371 and DT 7-9   ☐ State-Specific Test Prep Online Keyword: MT7 TestPrep

Teacher's Name _____ Class _____ Date _____

Lesson Plan 8-1
Perimeter & Area of Rectangles & Parallelograms pp. 388–392 Day _____

Objective Students find the perimeter and area of rectangles and parallelograms.

> **NCTM Standards:** Specify locations and describe spatial relationships using coordinate geometry and other representational systems; Use visualization, spatial reasoning, and geometric modeling to solve problems; Apply appropriate techniques, tools, and formulas to determine measurements.

Pacing
- ☐ 45-minute Classes: 1 day
- ☐ 90-minute Classes: 1/2 day
- ☐ Other_____

WARM UP
- ☐ Warm Up TE p. 388 and Daily Transparency 8-1
- ☐ Problem of the Day TE p. 388 and Daily Transparency 8-1
- ☐ Countdown to Testing Transparency Week 15

TEACH
- ☐ Lesson Presentation CD-ROM 8-1
- ☐ Alternate Opener, Explorations Transparency 8-1, TE p. 388, and Exploration 8-1
- ☐ Reaching All Learners TE p. 389
- ☐ Teaching Transparency 8-1
- ☐ *Hands-On Lab Activities* 8-1
- ☐ *Know-It Notebook* 8-1

PRACTICE AND APPLY
- ☐ Example 1: Average: 1–3, 9–11, 17, 18, 27–37 Advanced: 9–11, 17, 18, 27–37
- ☐ Example 2: Average: 1–7, 9–15, 17, 18, 23, 27–37 Advanced: 9–15, 17, 18, 21–24, 27–37
- ☐ Example 3: Average: 1–20, 23, 27–37 Advanced: 9–37

REACHING ALL LEARNERS – Differentiated Instruction for students with

Developing Knowledge	On-level Knowledge	Advanced Knowledge	English Language Development
☐ Multiple Representations TE p. 389	☐ Multiple Representations TE p. 389	☐ Multiple Representations TE p. 389	☐ Multiple Representations TE p. 389
☐ Practice A 8-1 CRB	☐ Practice B 8-1 CRB	☐ Practice C 8-1 CRB	☐ Practice A, B, or C 8-1 CRB
☐ Reteach 8-1 CRB	☐ Puzzles, Twisters & Teasers 8-1 CRB	☐ Challenge 8-1 CRB	☐ *Success for ELL* 8-1
☐ Homework Help Online Keyword: MT7 8-1	☐ Homework Help Online Keyword: MT7 8-1	☐ Homework Help Online Keyword: MT7 8-1	☐ Homework Help Online Keyword: MT7 8-1
☐ *Lesson Tutorial Video* 8-1	☐ *Lesson Tutorial Video* 8-1	☐ *Lesson Tutorial Video* 8-1	☐ *Lesson Tutorial Video* 8-1
☐ Reading Strategies 8-1 CRB	☐ Problem Solving 8-1 CRB	☐ Problem Solving 8-1 CRB	☐ Reading Strategies 8-1 CRB
☐ *Questioning Strategies* pp. 110–111			☐ Lesson Vocabulary SE p. 388
☐ *IDEA Works!* 8-1			☐ *Multilingual Glossary*

ASSESSMENT
- ☐ Lesson Quiz, TE p. 392 and DT 8-1   ☐ State-Specific Test Prep Online Keyword: MT7 TestPrep

Copyright © Holt, Rinehart and Winston.
All rights reserved.

Holt Mathematics

Teacher's Name _____ Class _____ Date _____

Lesson Plan 8-2
Perimeter and Area of Triangles and Trapezoids pp. 394–398 Day _____

Objective Students find the perimeter and area of triangles and trapezoids.

> **NCTM Standards:** Specify locations and describe spatial relationships using coordinate geometry and other representational systems; Use visualization, spatial reasoning, and geometric modeling to solve problems; Apply appropriate techniques, tools, and formulas to determine measurements.

Pacing
- ☐ 45-minute Classes: 1 day   ☐ 90-minute Classes: 1/2 day   ☐ Other _____

WARM UP
- ☐ Warm Up TE p. 394 and Daily Transparency 8-2
- ☐ Problem of the Day TE p. 394 and Daily Transparency 8-2
- ☐ Countdown to Testing Transparency Week 15

TEACH
- ☐ Lesson Presentation CD-ROM 8-2
- ☐ Alternate Opener, Explorations Transparency 8-2, TE p. 394, and Exploration 8-2
- ☐ Reaching All Learners TE p. 395
- ☐ Teaching Transparency 8-2
- ☐ *Hands-On Lab Activities* 8-2
- ☐ *Technology Lab Activities* 8-2
- ☐ *Know-It Notebook* 8-2

PRACTICE AND APPLY
- ☐ Example 1: Average: 1–3, 12–14, 31, 33, 37–43 Advanced: 12–14, 31, 33, 37–43
- ☐ Example 2: Average: 1–6, 12–17, 31, 33, 37–43 Advanced: 12–17, 27, 31, 33, 37–43
- ☐ Example 3: Average: 1–7, 12–18, 31, 33, 37–43 Advanced: 12–18, 27, 31, 33, 37–43
- ☐ Example 4: Average: 1–24, 29–33, 37–43 Advanced: 12–43

REACHING ALL LEARNERS – Differentiated Instruction for students with

Developing Knowledge	On-level Knowledge	Advanced Knowledge	English Language Development
☐ Critical Thinking TE p. 395	☐ Critical Thinking TE p. 395	☐ Critical Thinking TE p. 395	☐ Critical Thinking TE p. 395
☐ Practice A 8-2 CRB	☐ Practice B 8-2 CRB	☐ Practice C 8-2 CRB	☐ Practice A, B, or C 8-2 CRB
☐ Reteach 8-2 CRB	☐ Puzzles, Twisters & Teasers 8-2 CRB	☐ Challenge 8-2 CRB	☐ *Success for ELL* 8-2
☐ Homework Help Online Keyword: MT7 8-2	☐ Homework Help Online Keyword: MT7 8-2	☐ Homework Help Online Keyword: MT7 8-2	☐ Homework Help Online Keyword: MT7 8-2
☐ *Lesson Tutorial Video* 8-2	☐ *Lesson Tutorial Video* 8-2	☐ *Lesson Tutorial Video* 8-2	☐ *Lesson Tutorial Video* 8-2
☐ Reading Strategies 8-2 CRB	☐ Problem Solving 8-2 CRB	☐ Problem Solving 8-2 CRB	☐ Reading Strategies 8-2 CRB
☐ Questioning Strategies pp. 112–113			
☐ *IDEA Works!* 8-2			☐ *Multilingual Glossary*

ASSESSMENT
- ☐ Lesson Quiz, TE p. 398 and DT 8-2   ☐ State-Specific Test Prep Online Keyword: MT7 TestPrep

Holt Mathematics

Teacher's Name _____ Class _____ Date _____

Lesson Plan 8-3
Circles pp. 400–403   Day _____

Objective Students find the circumference and area of circles.

> **NCTM Standards:** Specify locations and describe spatial relationships using coordinate geometry and other representational systems; Use visualization, spatial reasoning, and geometric modeling to solve problems; Apply appropriate techniques, tools, and formulas to determine measurements.

Pacing
☐ 45-minute Classes: 1 day   ☐ 90-minute Classes: 1/2 day   ☐ Other _____

WARM UP
☐ Warm Up TE p. 400 and Daily Transparency 8-3
☐ Problem of the Day TE p. 400 and Daily Transparency 8-3
☐ Countdown to Testing Transparency Week 16

TEACH
☐ Lesson Presentation CD-ROM 8-3
☐ Alternate Opener, Explorations Transparency 8-3, TE p. 400, and Exploration 8-3
☐ Reaching All Learners TE p. 401
☐ Teaching Transparency 8-3
☐ *Hands-On Lab Activities* 8-3
☐ *Know-It Notebook* 8-3

PRACTICE AND APPLY
☐ Example 1: Average: 1, 2, 7, 8, 30–37 Advanced: 7, 8, 16–18, 30–37
☐ Example 2: Average: 1–4, 7–10, 13–15, 30–37 Advanced: 7–10, 13–21, 27, 30–37
☐ Example 3: Average: 1–5, 7–11, 13–15, 30–37 Advanced: 7–11, 13–23, 27–37
☐ Example 4: Average: 1–15, 24–26, 30–37 Advanced: 7–16, 21–37

REACHING ALL LEARNERS – Differentiated Instruction for students with

Developing Knowledge	On-level Knowledge	Advanced Knowledge	English Language Development
☐ Kinesthetic Experience TE p. 401	☐ Kinesthetic Experience TE p. 401	☐ Kinesthetic Experience TE p. 401	☐ Kinesthetic Experience TE p. 401
☐ Practice A 8-3 CRB	☐ Practice B 8-3 CRB	☐ Practice C 8-3 CRB	☐ Practice A, B, or C 8-3 CRB
☐ Reteach 8-3 CRB	☐ Puzzles, Twisters & Teasers 8-3 CRB	☐ Challenge 8-3 CRB	☐ *Success for ELL* 8-3
☐ Homework Help Online Keyword: MT7 8-3	☐ Homework Help Online Keyword: MT7 8-3	☐ Homework Help Online Keyword: MT7 8-3	☐ Homework Help Online Keyword: MT7 8-3
☐ *Lesson Tutorial Video* 8-3	☐ *Lesson Tutorial Video* 8-3	☐ *Lesson Tutorial Video* 8-3	☐ *Lesson Tutorial Video* 8-3
☐ Reading Strategies 8-3 CRB	☐ Problem Solving 8-3 CRB	☐ Problem Solving 8-3 CRB	☐ Reading Strategies 8-3 CRB
☐ *Questioning Strategies* pp. 114–115			☐ Lesson Vocabulary SE p. 400
☐ *IDEA Works!* 8-3			☐ *Multilingual Glossary*

ASSESSMENT
☐ Lesson Quiz, TE p. 403 and DT 8-3   ☐ State-Specific Test Prep Online Keyword: MT7 TestPrep

Teacher's Name _____ Class _____ Date _____

Lesson Plan 8-4
Drawing Three-Dimensional Figures pp. 408–411   Day _____

Objective Students draw and identify parts of three-dimensional figures.

> **NCTM Standards:** Use visualization, spatial reasoning, and geometric modeling to solve problems; Understand measurable attributes of objects and the units, systems, and processes of measurement.

Pacing
☐ 45-minute Classes: 1 day   ☐ 90-minute Classes: 1/2 day   ☐ Other_____

WARM UP
☐ Warm Up TE p. 408 and Daily Transparency 8-4
☐ Problem of the Day TE p. 408 and Daily Transparency 8-4
☐ Countdown to Testing Transparency Week 16

TEACH
☐ Lesson Presentation CD-ROM 8-4
☐ Alternate Opener, Explorations Transparency 8-4, TE p. 408, and Exploration 8-4
☐ Reaching All Learners TE p. 409
☐ Teaching Transparency 8-4
☐ *Hands-On Lab Activities* 8-4
☐ *Know-It Notebook* 8-4

PRACTICE AND APPLY
☐ Example 1: Average: 1, 4, 17–23 Advanced: 4, 17–23
☐ Example 2: Average: 1, 2, 4, 5, 10–12, 17–23 Advanced: 4, 5, 10–13, 15–23
☐ Example 3: Average: 1–12, 14, 17–23 Advanced: 4–23

REACHING ALL LEARNERS – Differentiated Instruction for students with

Developing Knowledge	On-level Knowledge	Advanced Knowledge	English Language Development
☐ Multiple Representations TE p. 409	☐ Multiple Representations TE p. 409	☐ Multiple Representations TE p. 409	☐ Multiple Representations TE p. 409
☐ Practice A 8-4 CRB	☐ Practice B 8-4 CRB	☐ Practice C 8-4 CRB	☐ Practice A, B, or C 8-4 CRB
☐ Reteach 8-4 CRB	☐ Puzzles, Twisters & Teasers 8-4 CRB	☐ Challenge 8-4 CRB	☐ *Success for ELL* 8-4
☐ Homework Help Online Keyword: MT7 8-4	☐ Homework Help Online Keyword: MT7 8-4	☐ Homework Help Online Keyword: MT7 8-4	☐ Homework Help Online Keyword: MT7 8-4
☐ *Lesson Tutorial Video* 8-4	☐ *Lesson Tutorial Video* 8-4	☐ *Lesson Tutorial Video* 8-4	☐ *Lesson Tutorial Video* 8-4
☐ Reading Strategies 8-4 CRB	☐ Problem Solving 8-4 CRB	☐ Problem Solving 8-4 CRB	☐ Reading Strategies 8-4 CRB
☐ *Questioning Strategies* pp. 116–117	☐ Visual TE p. 409	☐ Visual TE p. 409	☐ Lesson Vocabulary SE p. 408
☐ *IDEA Works!* 8-4			☐ *Multilingual Glossary*

ASSESSMENT
☐ Lesson Quiz, TE p. 411 and DT 8-4   ☐ State-Specific Test Prep Online Keyword: MT7 TestPrep

Teacher's Name _____ Class _____ Date _____

Lesson Plan 8-5
Volume of Prisms and Cylinders pp. 413–417   Day _____

Objective Students find the volume of prisms and cylinders.

> **NCTM Standards:** Analyze change in various contexts; Use visualization, spatial reasoning, and geometric modeling to solve problems; Apply appropriate techniques, tools, and formulas to determine measurements.

Pacing
☐ 45-minute Classes: 1 day   ☐ 90-minute Classes: 1/2 day   ☐ Other_____

WARM UP
☐ Warm Up TE p. 413 and Daily Transparency 8-5
☐ Problem of the Day TE p. 413 and Daily Transparency 8-5
☐ Countdown to Testing Transparency Week 16

TEACH
☐ Lesson Presentation CD-ROM 8-5
☐ Alternate Opener, Explorations Transparency 8-5, TE p. 413, and Exploration 8-5
☐ Reaching All Learners TE p. 414
☐ Teaching Transparency 8-5
☐ *Hands-On Lab Activities* 8-5
☐ *Technology Lab Activities* 8-5
☐ *Know-It Notebook* 8-5

PRACTICE AND APPLY
☐ Example 1: Average: 1–3, 7–9, 20–25 Advanced: 7–9, 17, 20–25
☐ Example 2: Average: 1–4, 7–10, 20–25 Advanced: 7–10, 17, 20–25
☐ Example 3: Average: 1–5, 7–11, 13, 20–25 Advanced: 7–11, 13–18, 20–25
☐ Example 4: Average: 1–13, 20–25 Advanced: 7–25

REACHING ALL LEARNERS – Differentiated Instruction for students with

Developing Knowledge	On-level Knowledge	Advanced Knowledge	English Language Development
☐ Modeling TE p. 414	☐ Modeling TE p. 414	☐ Modeling TE p. 414	☐ Modeling TE p. 414
☐ Practice A 8-5 CRB	☐ Practice B 8-5 CRB	☐ Practice C 8-5 CRB	☐ Practice A, B, or C 8-5 CRB
☐ Reteach 8-5 CRB	☐ Puzzles, Twisters & Teasers 8-5 CRB	☐ Challenge 8-5 CRB	☐ *Success for ELL* 8-5
☐ Homework Help Online Keyword: MT7 8-5	☐ Homework Help Online Keyword: MT7 8-5	☐ Homework Help Online Keyword: MT7 8-5	☐ Homework Help Online Keyword: MT7 8-5
☐ *Lesson Tutorial Video* 8-5	☐ *Lesson Tutorial Video* 8-5	☐ *Lesson Tutorial Video* 8-5	☐ *Lesson Tutorial Video* 8-5
☐ Reading Strategies 8-5 CRB	☐ Problem Solving 8-5 CRB	☐ Problem Solving 8-5 CRB	☐ Reading Strategies 8-5 CRB
☐ *Questioning Strategies* pp. 118–119			☐ Lesson Vocabulary SE p. 413
☐ *IDEA Works!* 8-5			☐ *Multilingual Glossary*

ASSESSMENT
☐ Lesson Quiz, TE p. 417 and DT 8-5   ☐ State-Specific Test Prep Online Keyword: MT7 TestPrep

Teacher's Name _____ Class _____ Date _____

Lesson Plan 8-6
Volume of Pyramids and Cones pp. 420–424   Day _____

Objective Students find the volume of pyramids and cones.

> **NCTM Standards:** Analyze change in various contexts; Use visualization, spatial reasoning, and geometric modeling to solve problems; Apply appropriate techniques, tools, and formulas to determine measurements.

Pacing
☐ 45-minute Classes: 1 day   ☐ 90-minute Classes: 1/2 day   ☐ Other_____

WARM UP
☐ Warm Up TE p. 420 and Daily Transparency 8-6
☐ Problem of the Day TE p. 420 and Daily Transparency 8-6
☐ Countdown to Testing Transparency Week 17

TEACH
☐ Lesson Presentation CD-ROM 8-6
☐ Alternate Opener, Explorations Transparency 8-6, TE p. 420, and Exploration 8-6
☐ Reaching All Learners TE p. 421
☐ Teaching Transparency 8-6
☐ *Hands-On Lab Activities* 8-6
☐ *Technology Lab Activities* 8-6
☐ *Know-It Notebook* 8-6

PRACTICE AND APPLY
☐ Example 1: Average: 1–6, 10–15, 19, 21, 29–36 Advanced: 10–15, 19–22, 28–36
☐ Example 2: Average: 1–7, 10–16, 19, 21, 27, 29–36 Advanced: 10–16, 19–22, 27–36
☐ Example 3: Average: 1–8, 10–17, 19–27 odd, 29–36 Advanced: 10–17, 19–36
☐ Example 4: Average: 1–19, 21–27 odd, 29–36 Advanced: 10–36

REACHING ALL LEARNERS – Differentiated Instruction for students with

Developing Knowledge	On-level Knowledge	Advanced Knowledge	English Language Development
☐ Kinesthetic Experience TE p. 421	☐ Kinesthetic Experience TE p. 421	☐ Kinesthetic Experience TE p. 421	☐ Kinesthetic Experience TE p. 421
☐ Practice A 8-6 CRB	☐ Practice B 8-6 CRB	☐ Practice C 8-6 CRB	☐ Practice A, B, or C 8-6 CRB
☐ Reteach 8-6 CRB	☐ Puzzles, Twisters & Teasers 8-6 CRB	☐ Challenge 8-6 CRB	☐ *Success for ELL* 8-6
☐ Homework Help Online Keyword: MT7 8-6	☐ Homework Help Online Keyword: MT7 8-6	☐ Homework Help Online Keyword: MT7 8-6	☐ Homework Help Online Keyword: MT7 8-6
☐ *Lesson Tutorial Video* 8-6	☐ *Lesson Tutorial Video* 8-6	☐ *Lesson Tutorial Video* 8-6	☐ *Lesson Tutorial Video* 8-6
☐ Reading Strategies 8-6 CRB	☐ Problem Solving 8-6 CRB	☐ Problem Solving 8-6 CRB	☐ Reading Strategies 8-6 CRB
☐ *Questioning Strategies* pp. 120–121	☐ Critical Thinking TE p. 421	☐ Critical Thinking TE p. 421	☐ *Lesson Vocabulary* SE p. 420
☐ *IDEA Works!* 8-6			☐ *Multilingual Glossary*

ASSESSMENT
☐ Lesson Quiz, TE p. 424 and DT 8-6   ☐ State-Specific Test Prep Online Keyword: MT7 TestPrep

Teacher's Name _____ Class _____ Date _____

Lesson Plan 8-7
Surface Area of Prisms and Cylinders pp. 427–430   Day _____

Objective Students find the surface area of prisms and cylinders.

> **NCTM Standards:** Analyze change in various contexts; Use visualization, spatial reasoning, and geometric modeling to solve problems; Apply appropriate techniques, tools, and formulas to determine measurements.

Pacing
☐ 45-minute Classes: 1 day   ☐ 90-minute Classes: 1/2 day   ☐ Other_____

WARM UP
☐ Warm Up TE p. 427 and Daily Transparency 8-7
☐ Problem of the Day TE p. 427 and Daily Transparency 8-7
☐ Countdown to Testing Transparency Week 17

TEACH
☐ Lesson Presentation CD-ROM 8-7
☐ Alternate Opener, Explorations Transparency 8-7, TE p. 427, and Exploration 8-7
☐ Reaching All Learners TE p. 428
☐ Teaching Transparency 8-7
☐ *Hands-On Lab Activities* 8-7
☐ *Know-It Notebook* 8-7

PRACTICE AND APPLY
☐ Example 1: Average: 1–3, 6–8, 11, 12, 21–29 Advanced: 6–8, 11–14, 18, 21–29
☐ Example 2: Average: 1–4, 6–9, 11, 12, 19, 21–29 Advanced: 6–9, 11–14, 18, 19, 21–29
☐ Example 3: Average: 1–12, 15, 16, 19, 21–29 Advanced: 6–29

REACHING ALL LEARNERS – Differentiated Instruction for students with

Developing Knowledge	On-level Knowledge	Advanced Knowledge	English Language Development
☐ Inclusion TE p. 428	☐ Cognitive Strategies TE p. 428	☐ Cognitive Strategies TE p. 428	☐ Cognitive Strategies TE p. 428
☐ Practice A 8-7 CRB	☐ Practice B 8-7 CRB	☐ Practice C 8-7 CRB	☐ Practice A, B, or C 8-7 CRB
☐ Reteach 8-7 CRB	☐ Puzzles, Twisters & Teasers 8-7 CRB	☐ Challenge 8-7 CRB	☐ *Success for ELL* 8-7
☐ Homework Help Online Keyword: MT7 8-7	☐ Homework Help Online Keyword: MT7 8-7	☐ Homework Help Online Keyword: MT7 8-7	☐ Homework Help Online Keyword: MT7 8-7
☐ *Lesson Tutorial Video* 8-7	☐ *Lesson Tutorial Video* 8-7	☐ *Lesson Tutorial Video* 8-7	☐ *Lesson Tutorial Video* 8-7
☐ Reading Strategies 8-7 CRB	☐ Problem Solving 8-7 CRB	☐ Problem Solving 8-7 CRB	☐ Reading Strategies 8-7 CRB
☐ Questioning Strategies pp. 122–123			☐ Lesson Vocabulary SE p. 428
☐ *IDEA Works!* 8-7			☐ *Multilingual Glossary*

ASSESSMENT
☐ Lesson Quiz, TE p. 430 and DT 8-7   ☐ State-Specific Test Prep Online Keyword: MT7 TestPrep

Teacher's Name _____ Class _____ Date _____

Lesson Plan 8-8
Surface Area of Pyramids and Cones pp. 432–435   Day _____

Objective Students find the surface area of pyramids and cones.

> **NCTM Standards:** Analyze change in various contexts; Use visualization, spatial reasoning, and geometric modeling to solve problems; Apply appropriate techniques, tools, and formulas to determine measurements.

Pacing
☐ 45-minute Classes: 1 day   ☐ 90-minute Classes: 1/2 day   ☐ Other_____

WARM UP
☐ Warm Up TE p. 432 and Daily Transparency 8-8
☐ Problem of the Day TE p. 432 and Daily Transparency 8-8
☐ Countdown to Testing Transparency Week 18

TEACH
☐ Lesson Presentation CD-ROM 8-8
☐ Alternate Opener, Explorations Transparency 8-8, TE p. 432, and Exploration 8-8
☐ Reaching All Learners TE p. 433
☐ Teaching Transparency 8-8
☐ *Hands-On Lab Activities* 8-8
☐ *Know-It Notebook* 8-8

PRACTICE AND APPLY
☐ Example 1: Average: 1–3, 6–8, 11, 12, 19–26 Advanced: 6–8, 11, 12, 17, 19–26
☐ Example 2: Average: 1–4, 6–9, 11, 12, 19–26 Advanced: 6–9, 11, 12, 17, 19–26
☐ Example 3: Average: 1–14, 19–26 Advanced: 6–26

REACHING ALL LEARNERS – Differentiated Instruction for students with

Developing Knowledge	On-level Knowledge	Advanced Knowledge	English Language Development
☐ Inclusion TE p. 433	☐ Modeling TE p. 433	☐ Modeling TE p. 433	☐ Modeling TE p. 433
☐ Practice A 8-8 CRB	☐ Practice B 8-8 CRB	☐ Practice C 8-8 CRB	☐ Practice A, B, or C 8-8 CRB
☐ Reteach 8-8 CRB	☐ Puzzles, Twisters & Teasers 8-8 CRB	☐ Challenge 8-8 CRB	☐ *Success for ELL* 8-8
☐ Homework Help Online Keyword: MT7 8-8	☐ Homework Help Online Keyword: MT7 8-8	☐ Homework Help Online Keyword: MT7 8-8	☐ Homework Help Online Keyword: MT7 8-8
☐ *Lesson Tutorial Video* 8-8	☐ *Lesson Tutorial Video* 8-8	☐ *Lesson Tutorial Video* 8-8	☐ *Lesson Tutorial Video* 8-8
☐ Reading Strategies 8-8 CRB	☐ Problem Solving 8-8 CRB	☐ Problem Solving 8-8 CRB	☐ Reading Strategies 8-8 CRB
☐ *Questioning Strategies* pp. 124–125			☐ Lesson Vocabulary SE p. 432
☐ *IDEA Works!* 8-8			☐ *Multilingual Glossary*

ASSESSMENT
☐ Lesson Quiz, TE p. 435 and DT 8-8   ☐ State-Specific Test Prep Online Keyword: MT7 TestPrep

Teacher's Name _____ Class _____ Date _____

Lesson Plan 8-9
Spheres pp. 436–439   Day _____

Objective Students find the volume and surface area of spheres.

> **NCTM Standards:** Use visualization, spatial reasoning, and geometric modeling to solve problems; Apply appropriate techniques, tools, and formulas to determine measurements.

Pacing
- ☐ 45-minute Classes: 1 day ☐ 90-minute Classes: 1/2 day ☐ Other_____

WARM UP
- ☐ Warm Up TE p. 436 and Daily Transparency 8-9
- ☐ Problem of the Day TE p. 436 and Daily Transparency 8-9
- ☐ Countdown to Testing Transparency Week 18

TEACH
- ☐ Lesson Presentation CD-ROM 8-9
- ☐ Alternate Opener, Explorations Transparency 8-9, TE p. 436, and Exploration 8-9
- ☐ Reaching All Learners TE p. 437
- ☐ Teaching Transparency 8-9
- ☐ *Hands-On Lab Activities* 8-9
- ☐ *Technology Lab Activities* 8-9
- ☐ *Know-It Notebook* 8-9

PRACTICE AND APPLY
- ☐ Example 1: Average: 1–4, 10–13, 25, 28–36 Advanced: 10–13, 24, 25, 27–36
- ☐ Example 2: Average: 1–8, 10–17, 19–22, 25, 26, 28–36 Advanced: 10–17, 19–22, 24–36
- ☐ Example 3: Average: 1–22, 25, 26, 28–36 Advanced: 10–36

REACHING ALL LEARNERS – Differentiated Instruction for students with

Developing Knowledge	On-level Knowledge	Advanced Knowledge	English Language Development
☐ Concrete Manipulatives TE p. 437	☐ Concrete Manipulatives TE p. 437	☐ Concrete Manipulatives TE p. 437	☐ Concrete Manipulatives TE p. 437
☐ Practice A 8-9 CRB	☐ Practice B 8-9 CRB	☐ Practice C 8-9 CRB	☐ Practice A, B, or C 8-9 CRB
☐ Reteach 8-9 CRB	☐ Puzzles, Twisters & Teasers 8-9 CRB	☐ Challenge 8-9 CRB	☐ *Success for ELL* 8-9
☐ Homework Help Online Keyword: MT7 8-9	☐ Homework Help Online Keyword: MT7 8-9	☐ Homework Help Online Keyword: MT7 8-9	☐ Homework Help Online Keyword: MT7 8-9
☐ *Lesson Tutorial Video* 8-9	☐ *Lesson Tutorial Video* 8-9	☐ *Lesson Tutorial Video* 8-9	☐ *Lesson Tutorial Video* 8-9
☐ Reading Strategies 8-9 CRB	☐ Problem Solving 8-9 CRB	☐ Problem Solving 8-9 CRB	☐ Reading Strategies 8-9 CRB
☐ *Questioning Strategies* pp. 126–127			☐ Lesson Vocabulary SE p. 436
☐ *IDEA Works!* 8-9			☐ *Multilingual Glossary*

ASSESSMENT
- ☐ Lesson Quiz, TE p. 439 and DT 8-9 ☐ State-Specific Test Prep Online Keyword: MT7 TestPrep

Copyright © Holt, Rinehart and Winston.
All rights reserved.

Holt Mathematics

Teacher's Name _____ Class _____ Date _____

Lesson Plan 8-10
Scaling Three-Dimensional Figures pp. 440–443   Day _____

Objective Students make scale models of solid figures.

> **NCTM Standards:** Use visualization, spatial reasoning, and geometric modeling to solve problems; Understand measurable attributes of objects and the units, systems, and processes of measurement; Apply appropriate techniques, tools, and formulas to determine measurements.

Pacing
☐ 45-minute Classes: 1 day   ☐ 90-minute Classes: 1/2 day   ☐ Other _____

WARM UP
☐ Warm Up TE p. 440 and Daily Transparency 8-10
☐ Problem of the Day TE p. 440 and Daily Transparency 8-10
☐ Countdown to Testing Transparency Week 18

TEACH
☐ Lesson Presentation CD-ROM 8-10
☐ Alternate Opener, Explorations Transparency 8-10, TE p. 440, and Exploration 8-10
☐ Reaching All Learners TE p. 441
☐ Teaching Transparency 8-10
☐ *Know-It Notebook* 8-10

PRACTICE AND APPLY
☐ Example 1: Average: 1–3, 6–8, 11–13, 22, 23, 25–34 Advanced: 6–8, 11–18, 22–34
☐ Example 2: Average: 1–4, 6–9, 11–13, 22, 23, 25–34 Advanced: 6–9, 11–20, 22–34
☐ Example 3: Average: 1–13, 22, 23, 25–34 Advanced: 6–10, 14–34

REACHING ALL LEARNERS – Differentiated Instruction for students with

Developing Knowledge	On-level Knowledge	Advanced Knowledge	English Language Development
☐ Concrete Manipulatives TE p. 441	☐ Concrete Manipulatives TE p. 441	☐ Concrete Manipulatives TE p. 441	☐ Concrete Manipulatives TE p. 441
☐ Practice A 8-10 CRB	☐ Practice B 8-10 CRB	☐ Practice C 8-10 CRB	☐ Practice A, B, or C 8-10 CRB
☐ Reteach 8-10 CRB	☐ Puzzles, Twisters & Teasers 8-10 CRB	☐ Challenge 8-10 CRB	☐ *Success for ELL* 8-10
☐ Homework Help Online Keyword: MT7 8-10	☐ Homework Help Online Keyword: MT7 8-10	☐ Homework Help Online Keyword: MT7 8-10	☐ Homework Help Online Keyword: MT7 8-10
☐ *Lesson Tutorial Video* 8-10	☐ *Lesson Tutorial Video* 8-10	☐ *Lesson Tutorial Video* 8-10	☐ *Lesson Tutorial Video* 8-10
☐ Reading Strategies 8-10 CRB	☐ Problem Solving 8-10 CRB	☐ Problem Solving 8-10 CRB	☐ Reading Strategies 8-10 CRB
☐ *Questioning Strategies* pp. 128–129			☐ Lesson Vocabulary SE p. 440
☐ *IDEA Works!* 8-10			☐ *Multilingual Glossary*

ASSESSMENT
☐ Lesson Quiz, TE p. 443 and DT 8-10   ☐ State-Specific Test Prep Online Keyword: MT7 TestPrep

Holt Mathematics

Teacher's Name _____ Class _____ Date _____

Lesson Plan 9-1
Samples and Surveys pp. 462–465 Day _____

Objective Students identify sampling methods and recognize biased samples.

> **NCTM Standards:** Formulate questions that can be addressed with data and collect, organize, and display relevant data to answer them.

Pacing
☐ 45-minute Classes: 1 day ☐ 90-minute Classes: 1/2 day ☐ Other _____

WARM UP
☐ Warm Up TE p. 462 and Daily Transparency 9-1
☐ Problem of the Day TE p. 462 and Daily Transparency 9-1
☐ Countdown to Testing Transparency Week 19

TEACH
☐ Lesson Presentation CD-ROM 9-1
☐ Alternate Opener, Explorations Transparency 9-1, TE p. 462, and Exploration 9-1
☐ Reaching All Learners TE p. 463
☐ Teaching Transparency 9-1
☐ *Know-It Notebook* 9-1

PRACTICE AND APPLY
☐ Example 1: Average: 1, 2, 4, 5, 8–12, 18, 21–29 Advanced: 4, 5, 8–12, 18, 20–29
☐ Example 2: Average: 1–16, 18, 21–29 Advanced: 4–29

REACHING ALL LEARNERS – Differentiated Instruction for students with

Developing Knowledge	On-level Knowledge	Advanced Knowledge	English Language Development
☐ Inclusion TE p. 463	☐ Cooperative Learning TE p. 463	☐ Cooperative Learning TE p. 463	☐ Cooperative Learning TE p. 463
☐ Practice A 9-1 CRB	☐ Practice B 9-1 CRB	☐ Practice C 9-1 CRB	☐ Practice A, B, or C 9-1 CRB
☐ Reteach 9-1 CRB	☐ Puzzles, Twisters & Teasers 9-1 CRB	☐ Challenge 9-1 CRB	☐ *Success for ELL* 9-1
☐ Homework Help Online Keyword: MT7 9-1	☐ Homework Help Online Keyword: MT7 9-1	☐ Homework Help Online Keyword: MT7 9-1	☐ Homework Help Online Keyword: MT7 9-1
☐ *Lesson Tutorial Video* 9-1	☐ *Lesson Tutorial Video* 9-1	☐ *Lesson Tutorial Video* 9-1	☐ *Lesson Tutorial Video* 9-1
☐ Reading Strategies 9-1 CRB	☐ Problem Solving 9-1 CRB	☐ Problem Solving 9-1 CRB	☐ Reading Strategies 9-1 CRB
☐ *Questioning Strategies* pp. 130–131			☐ Lesson Vocabulary SE p. 462
☐ *IDEA Works!* 9-1			☐ *Multilingual Glossary*

ASSESSMENT
☐ Lesson Quiz, TE p. 465 and DT 9-1 ☐ State-Specific Test Prep Online Keyword: MT7 TestPrep

Teacher's Name _____ Class _____ Date _____

Lesson Plan 9-2
Organizing Data pp. 467–471   Day _____

Objective Students organize data in tables and stem-and-leaf plots.

> **NCTM Standards:** Formulate questions that can be addressed with data and collect, organize, and display relevant data to answer them; Select and use appropriate statistical methods to analyze data.

Pacing
☐ 45-minute Classes: 1 day   ☐ 90-minute Classes: 1/2 day   ☐ Other_____

WARM UP
☐ Warm Up TE p. 467 and Daily Transparency 9-2
☐ Problem of the Day TE p. 467 and Daily Transparency 9-2
☐ Countdown to Testing Transparency Week 19

TEACH
☐ Lesson Presentation CD-ROM 9-2
☐ Alternate Opener, Explorations Transparency 9-2, TE p. 467, and Exploration 9-2
☐ Reaching All Learners TE p. 468
☐ *Technology Lab Activities* 9-2
☐ *Know-It Notebook* 9-2

PRACTICE AND APPLY
☐ Example 1: Average: 1, 6, 12, 20, 22–29 Advanced: 6, 12, 20, 22–29
☐ Example 2: Average: 1–3, 6–8, 12, 20, 22–29 Advanced: 6–8, 12, 20, 22–29
☐ Example 3: Average: 1–4, 6–9, 11, 12, 17–20, 22–29 Advanced: 6–9, 11, 12, 17–29
☐ Example 4: Average: 1–20, 22–29 Advanced: 6–29

REACHING ALL LEARNERS – Differentiated Instruction for students with

Developing Knowledge	On-level Knowledge	Advanced Knowledge	English Language Development
☐ Kinesthetic Experience TE p. 468	☐ Kinesthetic Experience TE p. 468	☐ Kinesthetic Experience TE p. 468	☐ Kinesthetic Experience TE p. 468
☐ Practice A 9-2 CRB	☐ Practice B 9-2 CRB	☐ Practice C 9-2 CRB	☐ Practice A, B, or C 9-2 CRB
☐ Reteach 9-2 CRB	☐ Puzzles, Twisters & Teasers 9-2 CRB	☐ Challenge 9-2 CRB	☐ *Success for ELL* 9-2
☐ Homework Help Online Keyword: MT7 9-2	☐ Homework Help Online Keyword: MT7 9-2	☐ Homework Help Online Keyword: MT7 9-2	☐ Homework Help Online Keyword: MT7 9-2
☐ *Lesson Tutorial Video* 9-2	☐ *Lesson Tutorial Video* 9-2	☐ *Lesson Tutorial Video* 9-2	☐ *Lesson Tutorial Video* 9-2
☐ Reading Strategies 9-2 CRB	☐ Problem Solving 9-2 CRB	☐ Problem Solving 9-2 CRB	☐ Reading Strategies 9-2 CRB
☐ *Questioning Strategies* pp. 132–133			☐ Lesson Vocabulary SE p. 467
☐ *IDEA Works!* 9-2			☐ *Multilingual Glossary*

ASSESSMENT
☐ Lesson Quiz, TE p. 471 and DT 9-2   ☐ State-Specific Test Prep Online Keyword: MT7 TestPrep

Teacher's Name _____ Class _____ Date _____

Lesson Plan 9-3
Measures of Central Tendency pp. 472–475   Day _____

Objective Students find appropriate measures of central tendency.

> **NCTM Standards:** Select and use appropriate statistical methods to analyze data.

Pacing
☐ 45-minute Classes: 1 day   ☐ 90-minute Classes: 1/2 day   ☐ Other_____

WARM UP
☐ Warm Up TE p. 472 and Daily Transparency 9-3
☐ Problem of the Day TE p. 472 and Daily Transparency 9-3
☐ Countdown to Testing Transparency Week 19

TEACH
☐ Lesson Presentation CD-ROM 9-3
☐ Alternate Opener, Explorations Transparency 9-3, TE p. 472, and Exploration 9-3
☐ Reaching All Learners TE p. 473
☐ Teaching Transparency 9-3
☐ *Technology Lab Activities* 9-3
☐ *Know-It Notebook* 9-3

PRACTICE AND APPLY
☐ Example 1: Average: 1–4, 8–11, 24–29 Advanced: 8–11, 21, 23–29
☐ Example 2: Average: 1–6, 8–13, 15–18, 24–29 Advanced: 8–13, 15–18, 21, 23–29
☐ Example 3: Average: 1–19, 24–29 Advanced: 8–29

REACHING ALL LEARNERS – Differentiated Instruction for students with

Developing Knowledge	On-level Knowledge	Advanced Knowledge	English Language Development
☐ Inclusion TE p. 473	☐ Home Connection TE p. 473	☐ Home Connection TE p. 473	☐ Home Connection TE p. 473
☐ Practice A 9-3 CRB	☐ Practice B 9-3 CRB	☐ Practice C 9-3 CRB	☐ Practice A, B, or C 9-3 CRB
☐ Reteach 9-3 CRB	☐ Puzzles, Twisters & Teasers 9-3 CRB	☐ Challenge 9-3 CRB	☐ *Success for ELL* 9-3
☐ Homework Help Online Keyword: MT7 9-3	☐ Homework Help Online Keyword: MT7 9-3	☐ Homework Help Online Keyword: MT7 9-3	☐ Homework Help Online Keyword: MT7 9-3
☐ *Lesson Tutorial Video* 9-3	☐ *Lesson Tutorial Video* 9-3	☐ *Lesson Tutorial Video* 9-3	☐ *Lesson Tutorial Video* 9-3
☐ Reading Strategies 9-3 CRB	☐ Problem Solving 9-3 CRB	☐ Problem Solving 9-3 CRB	☐ Reading Strategies 9-3 CRB
☐ *Questioning Strategies* pp. 134–135			☐ Lesson Vocabulary SE p. 472
☐ *IDEA Works!* 9-3			☐ *Multilingual Glossary*

ASSESSMENT
☐ Lesson Quiz, TE p. 475 and DT 9-3   ☐ State-Specific Test Prep Online Keyword: MT7 TestPrep

Teacher's Name _____ Class _____ Date _____

Lesson Plan 9-4
Variability pp. 476–480 Day _____

Objective Students find measures of variability.

> **NCTM Standards:** Select and use appropriate statistical methods to analyze data; Recognize and apply mathematics in contexts outside of mathematics.

Pacing
☐ 45-minute Classes: 1 day   ☐ 90-minute Classes: 1/2 day   ☐ Other_____

WARM UP
☐ Warm Up TE p. 476 and Daily Transparency 9-4
☐ Problem of the Day TE p. 476 and Daily Transparency 9-4
☐ Countdown to Testing Transparency Week 19

TEACH
☐ Lesson Presentation CD-ROM 9-4
☐ Alternate Opener, Explorations Transparency 9-4, TE p. 476, and Exploration 9-4
☐ Reaching All Learners TE p. 477
☐ Teaching Transparency 9-4
☐ *Know-It Notebook* 9-4

PRACTICE AND APPLY
☐ Example 1: Average: 1, 2, 7, 8, 13–16, 27–35 Advanced: 7, 8, 13–16, 27–35
☐ Example 2: Average: 1–4, 7–10, 13–19, 27–35 Advanced: 7–10, 14–20, 22, 25–35
☐ Example 3: Average: 1–19, 21, 27–35 Advanced: 7–12, 15–35

REACHING ALL LEARNERS – Differentiated Instruction for students with

Developing Knowledge	On-level Knowledge	Advanced Knowledge	English Language Development
☐ Cognitive Strategies TE p. 477	☐ Cognitive Strategies TE p. 477	☐ Cognitive Strategies TE p. 477	☐ Cognitive Strategies TE p. 477
☐ Practice A 9-4 CRB	☐ Practice B 9-4 CRB	☐ Practice C 9-4 CRB	☐ Practice A, B, or C 9-4 CRB
☐ Reteach 9-4 CRB	☐ Puzzles, Twisters & Teasers 9-4 CRB	☐ Challenge 9-4 CRB	☐ *Success for ELL* 9-4
☐ Homework Help Online Keyword: MT7 9-4	☐ Homework Help Online Keyword: MT7 9-4	☐ Homework Help Online Keyword: MT7 9-4	☐ Homework Help Online Keyword: MT7 9-4
☐ *Lesson Tutorial Video* 9-4	☐ *Lesson Tutorial Video* 9-4	☐ *Lesson Tutorial Video* 9-4	☐ *Lesson Tutorial Video* 9-4
☐ Reading Strategies 9-4 CRB	☐ Problem Solving 9-4 CRB	☐ Problem Solving 9-4 CRB	☐ Reading Strategies 9-4 CRB
☐ *Questioning Strategies* pp. 136–137			☐ Lesson Vocabulary SE p. 476
☐ *IDEA Works!* 9-4			☐ *Multilingual Glossary*

ASSESSMENT
☐ Lesson Quiz, TE p. 480 and DT 9-4   ☐ State-Specific Test Prep Online Keyword: MT7 TestPrep

Teacher's Name _____ Class _____ Date _____

Lesson Plan 9-5
Displaying Data pp. 485–488 Day _____

Objective Students display data in bar graphs, histograms, and line graphs.

> **NCTM Standards:** Use mathematical models to represent and understand quantitative relationships.

Pacing
- [ ] 45-minute Classes: 1 day
- [ ] 90-minute Classes: 1/2 day
- [ ] Other _____

WARM UP
- [ ] Warm Up TE p. 485 and Daily Transparency 9-5
- [ ] Problem of the Day TE p. 485 and Daily Transparency 9-5
- [ ] Countdown to Testing Transparency Week 20

TEACH
- [ ] Lesson Presentation CD-ROM 9-5
- [ ] Alternate Opener, Explorations Transparency 9-5, TE p. 485, and Exploration 9-5
- [ ] Reaching All Learners TE p. 486
- [ ] *Hands-On Lab Activities* 9-5
- [ ] *Know-It Notebook* 9-5

PRACTICE AND APPLY
- [ ] Example 1: Average: 1, 4, 7, 12–19 Advanced: 4, 7, 12–19
- [ ] Example 2: Average: 1, 2, 4, 5, 7, 8, 12–19 Advanced: 4, 5, 7–9, 11–19
- [ ] Example 3: Average: 1–8, 12–19 Advanced: 4–19

REACHING ALL LEARNERS – Differentiated Instruction for students with

Developing Knowledge	On-level Knowledge	Advanced Knowledge	English Language Development
☐ Multiple Representations TE p. 486	☐ Multiple Representations TE p. 486	☐ Multiple Representations TE p. 486	☐ Multiple Representations TE p. 486
☐ Practice A 9-5 CRB	☐ Practice B 9-5 CRB	☐ Practice C 9-5 CRB	☐ Practice A, B, or C 9-5 CRB
☐ Reteach 9-5 CRB	☐ Puzzles, Twisters & Teasers 9-5 CRB	☐ Challenge 9-5 CRB	☐ *Success for ELL* 9-5
☐ Homework Help Online Keyword: MT7 9-5	☐ Homework Help Online Keyword: MT7 9-5	☐ Homework Help Online Keyword: MT7 9-5	☐ Homework Help Online Keyword: MT7 9-5
☐ *Lesson Tutorial Video* 9-5	☐ *Lesson Tutorial Video* 9-5	☐ *Lesson Tutorial Video* 9-5	☐ *Lesson Tutorial Video* 9-5
☐ Reading Strategies 9-5 CRB	☐ Problem Solving 9-5 CRB	☐ Problem Solving 9-5 CRB	☐ Reading Strategies 9-5 CRB
☐ *Questioning Strategies* pp. 138–139	☐ Modeling TE p. 486	☐ Modeling TE p. 486	☐ Lesson Vocabulary SE p. 485
☐ *IDEA Works!* 9-5			☐ *Multilingual Glossary*

ASSESSMENT
- [ ] Lesson Quiz, TE p. 488 and DT 9-5
- [ ] State-Specific Test Prep Online Keyword: MT7 TestPrep

Teacher's Name _____ Class _____ Date _____

Lesson Plan 9-6
Misleading Graphs and Statistics pp. 490–493   Day _____

Objective Students recognize misleading graphs and statistics.

> **NCTM Standards:** Select and use appropriate statistical methods to analyze data; Recognize and apply mathematics in contexts outside of mathematics.

Pacing
☐ 45-minute Classes: 1 day   ☐ 90-minute Classes: 1/2 day   ☐ Other _____

WARM UP
☐ Warm Up TE p. 490 and Daily Transparency 9-6
☐ Problem of the Day TE p. 490 and Daily Transparency 9-6
☐ Countdown to Testing Transparency Week 20

TEACH
☐ Lesson Presentation CD-ROM 9-6
☐ Alternate Opener, Explorations Transparency 9-6, TE p. 490, and Exploration 9-6
☐ Reaching All Learners TE p. 491
☐ *Know-It Notebook* 9-6

PRACTICE AND APPLY
☐ Example 1: Average: 1, 2, 5, 6, 9, 10, 13–17 Advanced: 5, 6, 9–17
☐ Example 2: Average: 1–10, 13–17 Advanced: 5–17

REACHING ALL LEARNERS – Differentiated Instruction for students with

Developing Knowledge	On-level Knowledge	Advanced Knowledge	English Language Development
☐ Critical Thinking TE p. 491	☐ Critical Thinking TE p. 491	☐ Critical Thinking TE p. 491	☐ Critical Thinking TE p. 491
☐ Practice A 9-6 CRB	☐ Practice B 9-6 CRB	☐ Practice C 9-6 CRB	☐ Practice A, B, or C 9-6 CRB
☐ Reteach 9-6 CRB	☐ Puzzles, Twisters & Teasers 9-6 CRB	☐ Challenge 9-6 CRB	☐ *Success for ELL* 9-6
☐ Homework Help Online Keyword: MT7 9-6	☐ Homework Help Online Keyword: MT7 9-6	☐ Homework Help Online Keyword: MT7 9-6	☐ Homework Help Online Keyword: MT7 9-6
☐ *Lesson Tutorial Video* 9-6	☐ *Lesson Tutorial Video* 9-6	☐ *Lesson Tutorial Video* 9-6	☐ *Lesson Tutorial Video* 9-6
☐ Reading Strategies 9-6 CRB	☐ Problem Solving 9-6 CRB	☐ Problem Solving 9-6 CRB	☐ Reading Strategies 9-6 CRB
☐ *Questioning Strategies* pp. 140–141			
☐ *IDEA Works!* 9-6			☐ *Multilingual Glossary*

ASSESSMENT
☐ Lesson Quiz, TE p. 493 and DT 9-6   ☐ State-Specific Test Prep Online Keyword: MT7 TestPrep

Teacher's Name _____ Class _____ Date _____

Lesson Plan 9-7
Scatter Plots pp. 494–497   Day _____

Objective Students create and interpret scatter plots.

> **NCTM Standards:** Formulate questions that can be addressed with data and collect, organize, and display relevant data to answer them.

Pacing
- [ ] 45-minute Classes: 1 day   [ ] 90-minute Classes: 1/2 day   [ ] Other _____

WARM UP
- [ ] Warm Up TE p. 494 and Daily Transparency 9-7
- [ ] Problem of the Day TE p. 494 and Daily Transparency 9-7
- [ ] Countdown to Testing Transparency Week 21

TEACH
- [ ] Lesson Presentation CD-ROM 9-7
- [ ] Alternate Opener, Explorations Transparency 9-7, TE p. 494, and Exploration 9-7
- [ ] Reaching All Learners TE p. 495
- [ ] Teaching Transparency 9-7
- [ ] *Hands-On Lab Activities* 9-7
- [ ] *Technology Lab Activities* 9-7
- [ ] *Know-It Notebook* 9-7

PRACTICE AND APPLY
- [ ] Example 1: Average: 1, 5, 14–19 Advanced: 5, 14–19
- [ ] Example 2: Average: 1–3, 5–7, 11, 12, 14–19 Advanced: 5–7, 9–19
- [ ] Example 3: Average: 1–8, 11, 12, 14–19 Advanced: 5–19

REACHING ALL LEARNERS – Differentiated Instruction for students with

Developing Knowledge	On-level Knowledge	Advanced Knowledge	English Language Development
[ ] Curriculum Integration TE p. 495	[ ] Curriculum Integration TE p. 495	[ ] Curriculum Integration TE p. 495	[ ] Curriculum Integration TE p. 495
[ ] Practice A 9-7 CRB	[ ] Practice B 9-7 CRB	[ ] Practice C 9-7 CRB	[ ] Practice A, B, or C 9-7 CRB
[ ] Reteach 9-7 CRB	[ ] Puzzles, Twisters & Teasers 9-7 CRB	[ ] Challenge 9-7 CRB	[ ] *Success for ELL* 9-7
[ ] Homework Help Online Keyword: MT7 9-7	[ ] Homework Help Online Keyword: MT7 9-7	[ ] Homework Help Online Keyword: MT7 9-7	[ ] Homework Help Online Keyword: MT7 9-7
[ ] *Lesson Tutorial Video* 9-7	[ ] *Lesson Tutorial Video* 9-7	[ ] *Lesson Tutorial Video* 9-7	[ ] *Lesson Tutorial Video* 9-7
[ ] Reading Strategies 9-7 CRB	[ ] Problem Solving 9-7 CRB	[ ] Problem Solving 9-7 CRB	[ ] Reading Strategies 9-7 CRB
[ ] *Questioning Strategies* pp. 142–143			[ ] Lesson Vocabulary SE p. 494
[ ] *IDEA Works!* 9-7			[ ] *Multilingual Glossary*

ASSESSMENT
- [ ] Lesson Quiz, TE p. 497 and DT 9-7   [ ] State-Specific Test Prep Online Keyword: MT7 TestPrep

Teacher's Name _____ Class _____ Date _____

Lesson Plan 9-8
Choosing the Best Representation of Data pp. 500–503 Day _____

Objective Students will select the best representation for a set of data.

> **NCTM Standards:** Select and use appropriate statistical methods to analyze data; Recognize and apply mathematics in contexts outside of mathematics.

Pacing
☐ 45-minute Classes: 1 day ☐ 90-minute Classes: 1/2 day ☐ Other _____

WARM UP
☐ Warm Up TE p. 500 and Daily Transparency 9-8
☐ Problem of the Day TE p. 500 and Daily Transparency 9-8
☐ Countdown to Testing Transparency Week 21

TEACH
☐ Lesson Presentation CD-ROM 9-8
☐ Alternate Opener, Explorations Transparency 9-8, TE p. 500, and Exploration 9-8
☐ Reaching All Learners TE p. 501
☐ Teaching Transparency 9-8
☐ *Hands-On Lab Activities* 9-8
☐ *Technology Lab Activities* 9-8
☐ *Know-It Notebook* 9-8

PRACTICE AND APPLY
☐ Example 1: Average: 1, 3, 14–18 Advanced: 3, 12, 14–18
☐ Example 2: Average: 1–10, 14–18 Advanced: 3–18

REACHING ALL LEARNERS – Differentiated Instruction for students with

Developing Knowledge	On-level Knowledge	Advanced Knowledge	English Language Development
☐ Cooperative Learning TE p. 501	☐ Cooperative Learning TE p. 501	☐ Cooperative Learning TE p. 501	☐ Cooperative Learning TE p. 501
☐ Practice A 9-8 CRB	☐ Practice B 9-8 CRB	☐ Practice C 9-8 CRB	☐ Practice A, B, or C 9-8 CRB
☐ Reteach 9-8 CRB	☐ Puzzles, Twisters & Teasers 9-8 CRB	☐ Challenge 9-8 CRB	☐ *Success for ELL* 9-8
☐ Homework Help Online Keyword: MT7 9-8	☐ Homework Help Online Keyword: MT7 9-8	☐ Homework Help Online Keyword: MT7 9-8	☐ Homework Help Online Keyword: MT7 9-8
☐ *Lesson Tutorial Video* 9-8	☐ *Lesson Tutorial Video* 9-8	☐ *Lesson Tutorial Video* 9-8	☐ *Lesson Tutorial Video* 9-8
☐ Reading Strategies 9-8 CRB	☐ Problem Solving 9-8 CRB	☐ Problem Solving 9-8 CRB	☐ Reading Strategies 9-8 CRB
☐ *Questioning Strategies* pp. 144–145			
☐ *IDEA Works!* 9-8			☐ *Multilingual Glossary*

ASSESSMENT
☐ Lesson Quiz, TE p. 503 and DT 9-8 ☐ State-Specific Test Prep Online Keyword: MT7 TestPrep

Teacher's Name _____ Class _____ Date _____

Lesson Plan 10-1
Probability pp. 522–526   Day _____

Objective Students find the probability of an event by using the definition of probability.

> **NCTM Standards:** Develop and evaluate inferences and predictions that are based on data; Understand and apply basic concepts of probability.

Pacing
☐ 45-minute Classes: 1 day   ☐ 90-minute Classes: 1/2 day   ☐ Other_____

WARM UP
☐ Warm Up TE p. 522 and Daily Transparency 10-1
☐ Problem of the Day TE p. 522 and Daily Transparency 10-1
☐ Countdown to Testing Transparency Week 22

TEACH
☐ Lesson Presentation CD-ROM 10-1
☐ Alternate Opener, Explorations Transparency 10-1, TE p. 522, and Exploration 10-1
☐ Reaching All Learners TE p. 523
☐ Teaching Transparency 10-1
☐ *Hands-On Lab Activities* 10-1
☐ *Know-It Notebook* 10-1

PRACTICE AND APPLY
☐ Example 1: Average: 1, 5, 19–28   Advanced: 5, 15–28
☐ Example 2: Average: 1–3, 5–7, 9–12, 19–28   Advanced: 5–7, 11, 12, 15–28
☐ Example 3: Average: 1–13, 19–28   Advanced: 5–8, 11–28

REACHING ALL LEARNERS – Differentiated Instruction for students with

Developing Knowledge	On-level Knowledge	Advanced Knowledge	English Language Development
☐ Cognitive Strategies TE p. 523	☐ Cognitive Strategies TE p. 523	☐ Cognitive Strategies TE p. 523	☐ Cognitive Strategies TE p. 523
☐ Practice A 10-1 CRB	☐ Practice B 10-1 CRB	☐ Practice C 10-1 CRB	☐ Practice A, B, or C 10-1 CRB
☐ Reteach 10-1 CRB	☐ Puzzles, Twisters & Teasers 10-1 CRB	☐ Challenge 10-1 CRB	☐ *Success for ELL* 10-1
☐ Homework Help Online Keyword: MT7 10-1	☐ Homework Help Online Keyword: MT7 10-1	☐ Homework Help Online Keyword: MT7 10-1	☐ Homework Help Online Keyword: MT7 10-1
☐ *Lesson Tutorial Video* 10-1	☐ *Lesson Tutorial Video* 10-1	☐ *Lesson Tutorial Video* 10-1	☐ *Lesson Tutorial Video* 10-1
☐ Reading Strategies 10-1 CRB	☐ Problem Solving 10-1 CRB	☐ Problem Solving 10-1 CRB	☐ Reading Strategies 10-1 CRB
☐ *Questioning Strategies* pp. 146–147			☐ Lesson Vocabulary SE p. 522
☐ *IDEA Works!* 10-1			☐ *Multilingual Glossary*

ASSESSMENT
☐ Lesson Quiz, TE p. 526 and DT 10-1   ☐ State-Specific Test Prep Online Keyword: MT7 TestPrep

Holt Mathematics

Teacher's Name _____ Class _____ Date _____

Lesson Plan 10-2
Experimental Probability pp. 527–530   Day _____

Objective Students estimate probability using experimental methods.

> **NCTM Standards:** Develop and evaluate inferences and predictions that are based on data; Understand and apply basic concepts of probability.

Pacing
☐ 45-minute Classes: 1 day   ☐ 90-minute Classes: 1/2 day   ☐ Other _____

WARM UP
☐ Warm Up TE p. 527 and Daily Transparency 10-2
☐ Problem of the Day TE p. 527 and Daily Transparency 10-2
☐ Countdown to Testing Transparency Week 22

TEACH
☐ Lesson Presentation CD-ROM 10-2
☐ Alternate Opener, Explorations Transparency 10-2, TE p. 527, and Exploration 10-2
☐ Reaching All Learners TE p. 528
☐ *Know-It Notebook* 10-2

PRACTICE AND APPLY
☐ Example 1: Average: 1, 2, 5, 6, 18–25 Advanced: 5, 6, 18–25
☐ Example 2: Average: 1–16, 18–25 Advanced: 5–25

REACHING ALL LEARNERS – Differentiated Instruction for students with

Developing Knowledge	On-level Knowledge	Advanced Knowledge	English Language Development
☐ Concrete Manipulatives TE p. 528	☐ Concrete Manipulatives TE p. 528	☐ Concrete Manipulatives TE p. 528	☐ Concrete Manipulatives TE p. 528
☐ Practice A 10-2 CRB	☐ Practice B 10-2 CRB	☐ Practice C 10-2 CRB	☐ Practice A, B, or C 10-2 CRB
☐ Reteach 10-2 CRB	☐ Puzzles, Twisters & Teasers 10-2 CRB	☐ Challenge 10-2 CRB	☐ *Success for ELL* 10-2
☐ Homework Help Online Keyword: MT7 10-2	☐ Homework Help Online Keyword: MT7 10-2	☐ Homework Help Online Keyword: MT7 10-2	☐ Homework Help Online Keyword: MT7 10-2
☐ *Lesson Tutorial Video* 10-2	☐ *Lesson Tutorial Video* 10-2	☐ *Lesson Tutorial Video* 10-2	☐ *Lesson Tutorial Video* 10-2
☐ Reading Strategies 10-2 CRB	☐ Problem Solving 10-2 CRB	☐ Problem Solving 10-2 CRB	☐ Reading Strategies 10-2 CRB
☐ *Questioning Strategies* pp. 148–149			☐ Lesson Vocabulary SE p. 527
☐ *IDEA Works!* 10-2			☐ *Multilingual Glossary*

ASSESSMENT
☐ Lesson Quiz, TE p. 530 and DT 10-2   ☐ State-Specific Test Prep Online Keyword: MT7 TestPrep

Teacher's Name _____ Class _____ Date _____

Lesson Plan 10-3
Use a Simulation pp. 532–535   Day _____

Objective Students use a simulation to estimate probability.

> **NCTM Standards:** Develop and evaluate inferences and predictions that are based on data; Understand and apply basic concepts of probability.

Pacing
☐ 45-minute Classes: 1 day   ☐ 90-minute Classes: 1/2 day   ☐ Other _____

WARM UP
☐ Warm Up TE p. 532 and Daily Transparency 10-3
☐ Problem of the Day TE p. 532 and Daily Transparency 10-3
☐ Countdown to Testing Transparency Week 22

TEACH
☐ Lesson Presentation CD-ROM 10-3
☐ Alternate Opener, Explorations Transparency 10-3, TE p. 532, and Exploration 10-3
☐ Reaching All Learners TE p. 533
☐ Teaching Transparency 10-3
☐ *Hands-On Lab Activities* 10-3
☐ *Know-It Notebook* 10-3

PRACTICE AND APPLY
☐ Example 1: Average: 1–10, 13–20 Advanced: 5–20

REACHING ALL LEARNERS – Differentiated Instruction for students with

Developing Knowledge	On-level Knowledge	Advanced Knowledge	English Language Development
☐ Critical Thinking TE p. 533	☐ Critical Thinking TE p. 533	☐ Critical Thinking TE p. 533	☐ Critical Thinking TE p. 533
☐ Practice A 10-3 CRB	☐ Practice B 10-3 CRB	☐ Practice C 10-3 CRB	☐ Practice A, B, or C 10-3 CRB
☐ Reteach 10-3 CRB	☐ Puzzles, Twisters & Teasers 10-3 CRB	☐ Challenge 10-3 CRB	☐ *Success for ELL* 10-3
☐ Homework Help Online Keyword: MT7 10-3	☐ Homework Help Online Keyword: MT7 10-3	☐ Homework Help Online Keyword: MT7 10-3	☐ Homework Help Online Keyword: MT7 10-3
☐ *Lesson Tutorial Video* 10-3	☐ *Lesson Tutorial Video* 10-3	☐ *Lesson Tutorial Video* 10-3	☐ *Lesson Tutorial Video* 10-3
☐ Reading Strategies 10-3 CRB	☐ Problem Solving 10-3 CRB	☐ Problem Solving 10-3 CRB	☐ Reading Strategies 10-3 CRB
☐ *Questioning Strategies* pp. 150–151			☐ Lesson Vocabulary SE p. 532
☐ *IDEA Works!* 10-3			☐ *Multilingual Glossary*

ASSESSMENT
☐ Lesson Quiz, TE p. 535 and DT 10-3   ☐ State-Specific Test Prep Online Keyword: MT7 TestPrep

Teacher's Name _____ Class _____ Date _____

Lesson Plan 10-4
Theoretical Probability pp. 540–544   Day _____

Objective Students estimate probability using theoretical methods.

> **NCTM Standards:** Develop and evaluate inferences and predictions that are based on data; Understand and apply basic concepts of probability.

Pacing
☐ 45-minute Classes: 1 day   ☐ 90-minute Classes: 1/2 day   ☐ Other_____

WARM UP
☐ Warm Up TE p. 540 and Daily Transparency 10-4
☐ Problem of the Day TE p. 540 and Daily Transparency 10-4
☐ Countdown to Testing Transparency Week 22

TEACH
☐ Lesson Presentation CD-ROM 10-4
☐ Alternate Opener, Explorations Transparency 10-4, TE p. 540, and Exploration 10-4
☐ Reaching All Learners TE p. 541
☐ Teaching Transparency 10-4
☐ *Know-It Notebook* 10-4

PRACTICE AND APPLY
☐ Example 1: Average: 1, 2, 9–12, 28–34 Advanced: 9–12, 28–34
☐ Example 2: Average: 1–6, 9–16, 19–23, 28–34 Advanced: 9–16, 19–23, 28–34
☐ Example 3: Average: 1–7, 9–17, 19–23, 28–34 Advanced: 9–17, 19–23, 28–34
☐ Example 4: Average: 1–26, 28–34 Advanced: 9–34

REACHING ALL LEARNERS – Differentiated Instruction for students with

Developing Knowledge	On-level Knowledge	Advanced Knowledge	English Language Development
☐ Cooperative Learning TE p. 541	☐ Cooperative Learning TE p. 541	☐ Cooperative Learning TE p. 541	☐ Cooperative Learning TE p. 541
☐ Practice A 10-4 CRB	☐ Practice B 10-4 CRB	☐ Practice C 10-4 CRB	☐ Practice A, B, or C 10-4 CRB
☐ Reteach 10-4 CRB	☐ Puzzles, Twisters & Teasers 10-4 CRB	☐ Challenge 10-4 CRB	☐ *Success for ELL* 10-4
☐ Homework Help Online Keyword: MT7 10-4	☐ Homework Help Online Keyword: MT7 10-4	☐ Homework Help Online Keyword: MT7 10-4	☐ Homework Help Online Keyword: MT7 10-4
☐ *Lesson Tutorial Video* 10-4	☐ *Lesson Tutorial Video* 10-4	☐ *Lesson Tutorial Video* 10-4	☐ *Lesson Tutorial Video* 10-4
☐ Reading Strategies 10-4 CRB	☐ Problem Solving 10-4 CRB	☐ Problem Solving 10-4 CRB	☐ Reading Strategies 10-4 CRB
☐ *Questioning Strategies* pp. 152–153			☐ Lesson Vocabulary SE p. 540
☐ *IDEA Works!* 10-4			☐ *Multilingual Glossary*

ASSESSMENT
☐ Lesson Quiz, TE p. 544 and DT 10-4   ☐ State-Specific Test Prep Online Keyword: MT7 TestPrep

Teacher's Name _____ Class _____ Date _____

Lesson Plan 10-5
Independent and Dependent Events pp. 545–549   Day _____

Objective Students find the probabilities of independent and dependent events.

> **NCTM Standards:** Develop and evaluate inferences and predictions that are based on data; Understand and apply basic concepts of probability; Make and investigate mathematical conjectures.

Pacing
☐ 45-minute Classes: 1 day   ☐ 90-minute Classes: 1/2 day   ☐ Other _____

WARM UP
☐ Warm Up TE p. 545 and Daily Transparency 10-5
☐ Problem of the Day TE p. 545 and Daily Transparency 10-5
☐ Countdown to Testing Transparency Week 23

TEACH
☐ Lesson Presentation CD-ROM 10-5
☐ Alternate Opener, Explorations Transparency 10-5, TE p. 545, and Exploration 10-5
☐ Reaching All Learners TE p. 546
☐ Teaching Transparency 10-5
☐ *Hands-On Lab Activities* 10-5
☐ *Technology Lab Activities* 10-5
☐ *Know-It Notebook* 10-5

PRACTICE AND APPLY
☐ Example 1: Average: 1, 2, 6, 7, 20–24 Advanced: 6, 7, 20–24
☐ Example 2: Average: 1–3, 6–9, 13, 15, 20–24 Advanced: 6–9, 13–15, 20–24
☐ Example 3: Average: 1–13, 15, 16, 20–24 Advanced: 6–24

REACHING ALL LEARNERS – Differentiated Instruction for students with

Developing Knowledge	On-level Knowledge	Advanced Knowledge	English Language Development
☐ Inclusion TE p. 546	☐ Critical Thinking TE p. 546	☐ Critical Thinking TE p. 546	☐ Critical Thinking TE p. 546
☐ Practice A 10-5 CRB	☐ Practice B 10-5 CRB	☐ Practice C 10-5 CRB	☐ Practice A, B, or C 10-5 CRB
☐ Reteach 10-5 CRB	☐ Puzzles, Twisters & Teasers 10-5 CRB	☐ Challenge 10-5 CRB	☐ *Success for ELL* 10-5
☐ Homework Help Online Keyword: MT7 10-5	☐ Homework Help Online Keyword: MT7 10-5	☐ Homework Help Online Keyword: MT7 10-5	☐ Homework Help Online Keyword: MT7 10-5
☐ *Lesson Tutorial Video* 10-5	☐ *Lesson Tutorial Video* 10-5	☐ *Lesson Tutorial Video* 10-5	☐ *Lesson Tutorial Video* 10-5
☐ Reading Strategies 10-5 CRB	☐ Problem Solving 10-5 CRB	☐ Problem Solving 10-5 CRB	☐ Reading Strategies 10-5 CRB
☐ *Questioning Strategies* pp. 154–155			☐ Lesson Vocabulary SE p. 545
☐ *IDEA Works!* 10-5			☐ *Multilingual Glossary*

ASSESSMENT
☐ Lesson Quiz, TE p. 549 and DT 10-5   ☐ State-Specific Test Prep Online Keyword: MT7 TestPrep

Teacher's Name _____ Class _____ Date _____

Lesson Plan 10-6
Making Decisions and Predictions pp. 550–553   Day _____

Objective Students use probability to make decisions and predictions.

> **NCTM Standards:** Develop and evaluate inferences and predictions that are based on data; Understand and apply basic concepts of probability; Make and investigate mathematical conjectures.

Pacing
☐ 45-minute Classes: 1 day   ☐ 90-minute Classes: 1/2 day   ☐ Other_____

WARM UP
☐ Warm Up TE p. 550 and Daily Transparency 10-6
☐ Problem of the Day TE p. 550 and Daily Transparency 10-6
☐ Countdown to Testing Transparency Week 23

TEACH
☐ Lesson Presentation CD-ROM 10-6
☐ Alternate Opener, Explorations Transparency 10-6, TE p. 550, and Exploration 10-6
☐ Reaching All Learners TE p. 551
☐ *Technology Lab Activities* 10-6
☐ *Know-It Notebook* 10-6

PRACTICE AND APPLY
☐ Example 1: Average: 1–3, 6, 7, 10–14, 16, 17, 21–27   Advanced: 6, 7, 10–14, 16–27
☐ Example 2: Average: 1–14, 16, 17, 21–27   Advanced: 6–27

REACHING ALL LEARNERS – Differentiated Instruction for students with

Developing Knowledge	On-level Knowledge	Advanced Knowledge	English Language Development
☐ Kinesthetic Experience TE p. 551	☐ Kinesthetic Experience TE p. 551	☐ Kinesthetic Experience TE p. 551	☐ Kinesthetic Experience TE p. 551
☐ Practice A 10-6 CRB	☐ Practice B 10-6 CRB	☐ Practice C 10-6 CRB	☐ Practice A, B, or C 10-6 CRB
☐ Reteach 10-6 CRB	☐ Puzzles, Twisters & Teasers 10-6 CRB	☐ Challenge 10-6 CRB	☐ *Success for ELL* 10-6
☐ Homework Help Online Keyword: MT7 10-6	☐ Homework Help Online Keyword: MT7 10-6	☐ Homework Help Online Keyword: MT7 10-6	☐ Homework Help Online Keyword: MT7 10-6
☐ *Lesson Tutorial Video* 10-6	☐ *Lesson Tutorial Video* 10-6	☐ *Lesson Tutorial Video* 10-6	☐ *Lesson Tutorial Video* 10-6
☐ Reading Strategies 10-6 CRB	☐ Problem Solving 10-6 CRB	☐ Problem Solving 10-6 CRB	☐ Reading Strategies 10-6 CRB
☐ *Questioning Strategies* pp. 156–157			
☐ *IDEA Works!* 10-6			☐ *Multilingual Glossary*

ASSESSMENT
☐ Lesson Quiz, TE p. 553 and DT 10-6   ☐ State-Specific Test Prep Online Keyword: MT7 TestPrep

Holt Mathematics

Teacher's Name _____ Class _____ Date _____

Lesson Plan 10-7
Odds pp. 554–557   Day _____

Objective Students convert between probabilities and odds.

> **NCTM Standards:** Develop and evaluate inferences and predictions that are based on data; Understand and apply basic concepts of probability.

Pacing
☐ 45-minute Classes: 1 day   ☐ 90-minute Classes: 1/2 day   ☐ Other_____

WARM UP
☐ Warm Up TE p. 554 and Daily Transparency 10-7
☐ Problem of the Day TE p. 554 and Daily Transparency 10-7
☐ Countdown to Testing Transparency Week 23

TEACH
☐ Lesson Presentation CD-ROM 10-7
☐ Alternate Opener, Explorations Transparency 10-7, TE p. 554, and Exploration 10-7
☐ Reaching All Learners TE p. 555
☐ Teaching Transparency 10-7
☐ *Know-It Notebook* 10-7

PRACTICE AND APPLY
☐ Example 1: Average: 1, 2, 7, 8, 13–18, 27–31 Advanced: 7, 8, 13–18, 21, 24–31
☐ Example 2: Average: 1–4, 7–10, 13–18, 22, 27–31 Advanced: 7–10, 13–18, 21–31
☐ Example 3: Average: 1–20, 22, 27–31 Advanced: 7–31

REACHING ALL LEARNERS – Differentiated Instruction for students with

Developing Knowledge	On-level Knowledge	Advanced Knowledge	English Language Development
☐ Graphic Organizers TE p. 555	☐ Graphic Organizers TE p. 555	☐ Graphic Organizers TE p. 555	☐ Graphic Organizers TE p. 555
☐ Practice A 10-7 CRB	☐ Practice B 10-7 CRB	☐ Practice C 10-7 CRB	☐ Practice A, B, or C 10-7 CRB
☐ Reteach 10-7 CRB	☐ Puzzles, Twisters & Teasers 10-7 CRB	☐ Challenge 10-7 CRB	☐ *Success for ELL* 10-7
☐ Homework Help Online Keyword: MT7 10-7	☐ Homework Help Online Keyword: MT7 10-7	☐ Homework Help Online Keyword: MT7 10-7	☐ Homework Help Online Keyword: MT7 10-7
☐ *Lesson Tutorial Video* 10-7	☐ *Lesson Tutorial Video* 10-7	☐ *Lesson Tutorial Video* 10-7	☐ *Lesson Tutorial Video* 10-7
☐ Reading Strategies 10-7 CRB	☐ Problem Solving 10-7 CRB	☐ Problem Solving 10-7 CRB	☐ Reading Strategies 10-7 CRB
☐ *Questioning Strategies* pp. 158–159	☐ Cognitive Strategies TE p. 555	☐ Cognitive Strategies TE p. 555	☐ Lesson Vocabulary SE p. 554
☐ *IDEA Works!* 10-7			☐ *Multilingual Glossary*

ASSESSMENT
☐ Lesson Quiz, TE p. 557 and DT 10-7   ☐ State-Specific Test Prep Online Keyword: MT7 TestPrep

Teacher's Name _____ Class _____ Date _____

Lesson Plan 10-8
Counting Principles pp. 558–562   Day _____

Objective Students find the number of possible outcomes in an experiment.

> **NCTM Standards:** Develop and evaluate inferences and predictions that are based on data; Understand and apply basic concepts of probability.

Pacing
☐ 45-minute Classes: 1 day   ☐ 90-minute Classes: 1/2 day   ☐ Other_____

WARM UP
☐ Warm Up TE p. 558 and Daily Transparency 10-8
☐ Problem of the Day TE p. 558 and Daily Transparency 10-8
☐ Countdown to Testing Transparency Week 23

TEACH
☐ Lesson Presentation CD-ROM 10-8
☐ Alternate Opener, Explorations Transparency 10-8, TE p. 558, and Exploration 10-8
☐ Reaching All Learners TE p. 559
☐ Teaching Transparency 10-8
☐ *Know-It Notebook* 10-8

PRACTICE AND APPLY
☐ Example 1: Average: 1–3, 6–8, 16, 17, 19, 20, 24–30 Advanced: 6–8, 12–20, 23–30
☐ Example 2: Average: 1–4, 6–10, 12–21, 24–30 Advanced: 6–10, 12–30
☐ Example 3: Average: 1–21, 24–30 Advanced: 6–30

REACHING ALL LEARNERS – Differentiated Instruction for students with

Developing Knowledge	On-level Knowledge	Advanced Knowledge	English Language Development
☐ Critical Thinking TE p. 559	☐ Critical Thinking TE p. 559	☐ Critical Thinking TE p. 559	☐ Critical Thinking TE p. 559
☐ Practice A 10-8 CRB	☐ Practice B 10-8 CRB	☐ Practice C 10-8 CRB	☐ Practice A, B, or C 10-8 CRB
☐ Reteach 10-8 CRB	☐ Puzzles, Twisters & Teasers 10-8 CRB	☐ Challenge 10-8 CRB	☐ *Success for ELL* 10-8
☐ Homework Help Online Keyword: MT7 10-8	☐ Homework Help Online Keyword: MT7 10-8	☐ Homework Help Online Keyword: MT7 10-8	☐ Homework Help Online Keyword: MT7 10-8
☐ *Lesson Tutorial Video* 10-8	☐ *Lesson Tutorial Video* 10-8	☐ *Lesson Tutorial Video* 10-8	☐ *Lesson Tutorial Video* 10-8
☐ Reading Strategies 10-8 CRB	☐ Problem Solving 10-8 CRB	☐ Problem Solving 10-8 CRB	☐ Reading Strategies 10-8 CRB
☐ *Questioning Strategies* pp. 160–161			☐ Lesson Vocabulary SE p. 558
☐ *IDEA Works!* 10-8			☐ *Multilingual Glossary*

ASSESSMENT
☐ Lesson Quiz, TE p. 562 and DT 10-8   ☐ State-Specific Test Prep Online Keyword: MT7 TestPrep

Teacher's Name _____ Class _____ Date _____

Lesson Plan 10-9

Permutations and Combinations pp. 563–567   Day _____

Objective Students find permutations and combinations.

> **NCTM Standards:** Compute fluently and make reasonable estimates.

Pacing
☐ 45-minute Classes: 1 day   ☐ 90-minute Classes: 1/2 day   ☐ Other_____

WARM UP
☐ Warm Up TE p. 563 and Daily Transparency 10-9
☐ Problem of the Day TE p. 563 and Daily Transparency 10-9
☐ Countdown to Testing Transparency Week 23

TEACH
☐ Lesson Presentation CD-ROM 10-9
☐ Alternate Opener, Explorations Transparency 10-9, TE p. 563, and Exploration 10-9
☐ Reaching All Learners TE p. 564
☐ Teaching Transparency 10-9
☐ *Hands-On Lab Activities* 10-9
☐ *Know-It Notebook* 10-9

PRACTICE AND APPLY
☐ Example 1: Average: 1–4, 9–12, 17, 18, 23, 41–45 Advanced: 9–12, 17, 18, 23, 26, 41–45
☐ Example 2: Average: 1–6, 9–14, 17–19, 23, 24, 33, 36, 41–45 Advanced: 9–14, 17–19, 24, 29, 30, 32, 33, 36, 41–45
☐ Example 3: Average: 1–24, 33–37, 41–45 Advanced: 9–16, 22–45

REACHING ALL LEARNERS – Differentiated Instruction for students with

Developing Knowledge	On-level Knowledge	Advanced Knowledge	English Language Development
☐ Critical Thinking TE p. 564	☐ Critical Thinking TE p. 564	☐ Critical Thinking TE p. 564	☐ Critical Thinking TE p. 564
☐ Practice A 10-9 CRB	☐ Practice B 10-9 CRB	☐ Practice C 10-9 CRB	☐ Practice A, B, or C 10-9 CRB
☐ Reteach 10-9 CRB	☐ Puzzles, Twisters & Teasers 10-9 CRB	☐ Challenge 10-9 CRB	☐ *Success for ELL* 10-9
☐ Homework Help Online Keyword: MT7 10-9	☐ Homework Help Online Keyword: MT7 10-9	☐ Homework Help Online Keyword: MT7 10-9	☐ Homework Help Online Keyword: MT7 10-9
☐ *Lesson Tutorial Video* 10-9	☐ *Lesson Tutorial Video* 10-9	☐ *Lesson Tutorial Video* 10-9	☐ *Lesson Tutorial Video* 10-9
☐ Reading Strategies 10-9 CRB	☐ Problem Solving 10-9 CRB	☐ Problem Solving 10-9 CRB	☐ Reading Strategies 10-9 CRB
☐ *Questioning Strategies* pp. 162–163			☐ Lesson Vocabulary SE p. 563
☐ *IDEA Works!* 10-9			☐ *Multilingual Glossary*

ASSESSMENT
☐ Lesson Quiz, TE p. 567 and DT 10-9   ☐ State-Specific Test Prep Online Keyword: MT7 TestPrep

Teacher's Name _____ Class _____ Date _____

Lesson Plan 11-1
Simplifying Algebraic Expressions pp. 584–587   Day _____

Objective Students combine like terms in an expression.

> **NCTM Standards:** Represent and analyze mathematical situations and structures using algebraic symbols; Understand how mathematical ideas interconnect and build on one another to produce a coherent whole.

Pacing
☐ 45-minute Classes: 1 day   ☐ 90-minute Classes: 1/2 day   ☐ Other_____

WARM UP
☐ Warm Up TE p. 584 and Daily Transparency 11-1
☐ Problem of the Day TE p. 584 and Daily Transparency 11-1
☐ Countdown to Testing Transparency Week 24

TEACH
☐ Lesson Presentation CD-ROM 11-1
☐ Alternate Opener, Explorations Transparency 11-1, TE p. 584, and Exploration 11-1
☐ Reaching All Learners TE p. 585
☐ *Hands-On Lab Activities* 11-1
☐ *Know-It Notebook* 11-1

PRACTICE AND APPLY
☐ Example 1: Average: 1–6, 19–27, 52, 58–65 Advanced: 19–27, 52, 55, 56, 58–65
☐ Example 2: Average: 1–12, 19–33, 52, 58–65 Advanced: 19–33, 52, 53, 55, 56, 58–65
☐ Example 3: Average: 1–15, 19–39, 47–49, 52, 58–65 Advanced: 19–39, 46–49, 52, 53, 55, 56, 58–65
☐ Example 4: Average: 1–45, 47–52, 58–65 Advanced: 19–65

REACHING ALL LEARNERS – Differentiated Instruction for students with

Developing Knowledge	On-level Knowledge	Advanced Knowledge	English Language Development
☐ Concrete Manipulatives TE p. 585	☐ Concrete Manipulatives TE p. 585	☐ Concrete Manipulatives TE p. 585	☐ Concrete Manipulatives TE p. 585
☐ Practice A 11-1 CRB	☐ Practice B 11-1 CRB	☐ Practice C 11-1 CRB	☐ Practice A, B, or C 11-1 CRB
☐ Reteach 11-1 CRB	☐ Puzzles, Twisters & Teasers 11-1 CRB	☐ Challenge 11-1 CRB	☐ *Success for ELL* 11-1
☐ Homework Help Online Keyword: MT7 11-1	☐ Homework Help Online Keyword: MT7 11-1	☐ Homework Help Online Keyword: MT7 11-1	☐ Homework Help Online Keyword: MT7 11-1
☐ *Lesson Tutorial Video* 11-1	☐ *Lesson Tutorial Video* 11-1	☐ *Lesson Tutorial Video* 11-1	☐ *Lesson Tutorial Video* 11-1
☐ Reading Strategies 11-1 CRB	☐ Problem Solving 11-1 CRB	☐ Problem Solving 11-1 CRB	☐ Reading Strategies 11-1 CRB
☐ *Questioning Strategies* pp. 164–165	☐ Visual TE p. 585	☐ Visual TE p. 585	☐ Lesson Vocabulary SE p. 584
☐ *IDEA Works!* 11-1			☐ *Multilingual Glossary*

ASSESSMENT
☐ Lesson Quiz, TE p. 587 and DT 11-1   ☐ State-Specific Test Prep Online Keyword: MT7 TestPrep

Teacher's Name _____ Class _____ Date _____

Lesson Plan 11-2
Solving Multi-Step Equations pp. 588–591   Day _____

Objective Students solve multi-step equations.

> **NCTM Standards:** Understand meanings of operations and how they relate to one another; Use mathematical models to represent and understand quantitative relationships.

Pacing
☐ 45-minute Classes: 1 day   ☐ 90-minute Classes: 1/2 day   ☐ Other _____

WARM UP
☐ Warm Up TE p. 588 and Daily Transparency 11-2
☐ Problem of the Day TE p. 588 and Daily Transparency 11-2
☐ Countdown to Testing Transparency Week 24

TEACH
☐ Lesson Presentation CD-ROM 11-2
☐ Alternate Opener, Explorations Transparency 11-2, TE p. 588, and Exploration 11-2
☐ Reaching All Learners TE p. 589
☐ *Technology Lab Activities* 11-2
☐ *Know-It Notebook* 11-2

PRACTICE AND APPLY
☐ Example 1: Average: 1–6, 12–17, 26, 27, 41–47 Advanced: 12–17, 26, 27, 29, 31, 32, 35, 39, 41–47
☐ Example 2: Average: 1–10, 12–23, 25–28, 41–47 Advanced: 12–23, 28–32, 35, 38–47
☐ Example 3: Average: 1–28, 33, 34, 41–47 Advanced: 12–24, 28–47

REACHING ALL LEARNERS – Differentiated Instruction for students with

Developing Knowledge	On-level Knowledge	Advanced Knowledge	English Language Development
☐ Inclusion TE p. 589	☐ Cooperative Learning TE p. 589	☐ Cooperative Learning TE p. 589	☐ Cooperative Learning TE p. 589
☐ Practice A 11-2 CRB	☐ Practice B 11-2 CRB	☐ Practice C 11-2 CRB	☐ Practice A, B, or C 11-2 CRB
☐ Reteach 11-2 CRB	☐ Puzzles, Twisters & Teasers 11-2 CRB	☐ Challenge 11-2 CRB	☐ *Success for ELL* 11-2
☐ Homework Help Online Keyword: MT7 11-2	☐ Homework Help Online Keyword: MT7 11-2	☐ Homework Help Online Keyword: MT7 11-2	☐ Homework Help Online Keyword: MT7 11-2
☐ *Lesson Tutorial Video* 11-2	☐ *Lesson Tutorial Video* 11-2	☐ *Lesson Tutorial Video* 11-2	☐ *Lesson Tutorial Video* 11-2
☐ Reading Strategies 11-2 CRB	☐ Problem Solving 11-2 CRB	☐ Problem Solving 11-2 CRB	☐ Reading Strategies 11-2 CRB
☐ *Questioning Strategies* pp. 166–167			
☐ *IDEA Works!* 11-2			☐ *Multilingual Glossary*

ASSESSMENT
☐ Lesson Quiz, TE p. 591 and DT 11-2   ☐ State-Specific Test Prep Online Keyword: MT7 TestPrep

Teacher's Name _____ Class _____ Date _____

Lesson Plan 11-3
Solving Equations with Variables on Both Sides pp. 593–597   Day _____

Objective Students solve equations with variables on both sides of the equal sign.

> **NCTM Standards:** Use mathematical models to represent and understand quantitative relationships.

Pacing
☐ 45-minute Classes: 1 day   ☐ 90-minute Classes: 1/2 day   ☐ Other_____

WARM UP
☐ Warm Up TE p. 593 and Daily Transparency 11-3
☐ Problem of the Day TE p. 593 and Daily Transparency 11-3
☐ Countdown to Testing Transparency Week 24

TEACH
☐ Lesson Presentation CD-ROM 11-3
☐ Alternate Opener, Explorations Transparency 11-3, TE p. 593, and Exploration 11-3
☐ Reaching All Learners TE p. 594
☐ *Technology Lab Activities* 11-3
☐ *Know-It Notebook* 11-3

PRACTICE AND APPLY
☐ Example 1: Average: 1–6, 13–18, 25, 26, 39–47 Advanced: 13–18, 25, 26, 39–47
☐ Example 2: Average: 1–10, 13–22, 25–28, 39–47 Advanced: 13–22, 27–32, 36, 38–47
☐ Example 3: Average: 1–11, 13–23, 25–28, 35, 39–47 Advanced: 13–23, 27–33, 35, 36, 38–47
☐ Example 4: Average: 1–28, 34, 35, 39–47 Advanced: 13–47

REACHING ALL LEARNERS – Differentiated Instruction for students with

Developing Knowledge	On-level Knowledge	Advanced Knowledge	English Language Development
☐ Kinesthetic Experience TE p. 594	☐ Kinesthetic Experience TE p. 594	☐ Kinesthetic Experience TE p. 594	☐ Kinesthetic Experience TE p. 594
☐ Practice A 11-3 CRB	☐ Practice B 11-3 CRB	☐ Practice C 11-3 CRB	☐ Practice A, B, or C 11-3 CRB
☐ Reteach 11-3 CRB	☐ Puzzles, Twisters & Teasers 11-3 CRB	☐ Challenge 11-3 CRB	☐ *Success for ELL* 11-3
☐ Homework Help Online Keyword: MT7 11-3	☐ Homework Help Online Keyword: MT7 11-3	☐ Homework Help Online Keyword: MT7 11-3	☐ Homework Help Online Keyword: MT7 11-3
☐ *Lesson Tutorial Video* 11-3	☐ *Lesson Tutorial Video* 11-3	☐ *Lesson Tutorial Video* 11-3	☐ *Lesson Tutorial Video* 11-3
☐ Reading Strategies 11-3 CRB	☐ Problem Solving 11-3 CRB	☐ Problem Solving 11-3 CRB	☐ Reading Strategies 11-3 CRB
☐ *Questioning Strategies* pp. 168–169			
☐ *IDEA Works!* 11-3			☐ *Multilingual Glossary*

ASSESSMENT
☐ Lesson Quiz, TE p. 597 and DT 11-3   ☐ State-Specific Test Prep Online Keyword: MT7 TestPrep

Teacher's Name _____ Class _____ Date _____

Lesson Plan 11-4
Solving Inequalities by Multiplying or Dividing pp. 600–603   Day _____

Objective Students solve and graph inequalities by using multiplication or division.

> **NCTM Standards:** Compute fluently and make reasonable estimates.

Pacing
☐ 45-minute Classes: 1 day   ☐ 90-minute Classes: 1/2 day   ☐ Other_____

WARM UP
☐ Warm Up TE p. 600 and Daily Transparency 11-4
☐ Problem of the Day TE p. 600 and Daily Transparency 11-4
☐ Countdown to Testing Transparency Week 24

TEACH
☐ Lesson Presentation CD-ROM 11-4
☐ Alternate Opener, Explorations Transparency 11-4, TE p. 600, and Exploration 11-4
☐ Reaching All Learners TE p. 601
☐ *Know-It Notebook* 11-4

PRACTICE AND APPLY
☐ Example 1: Average: 1–8, 10–17, 19–22, 33–34, 38–43 Advanced: 10–17, 22–30, 33–35, 38–43
☐ Example 2: Average: 1–22, 31–34, 38–43 Advanced: 10–18, 22–43

REACHING ALL LEARNERS – Differentiated Instruction for students with

Developing Knowledge	On-level Knowledge	Advanced Knowledge	English Language Development
☐ Visual Cues TE p. 601	☐ Visual Cues TE p. 601	☐ Visual Cues TE p. 601	☐ Visual Cues TE p. 601
☐ Practice A 11-4 CRB	☐ Practice B 11-4 CRB	☐ Practice C 11-4 CRB	☐ Practice A, B, or C 11-4 CRB
☐ Reteach 11-4 CRB	☐ Puzzles, Twisters & Teasers 11-4 CRB	☐ Challenge 11-4 CRB	☐ *Success for ELL* 11-4
☐ Homework Help Online Keyword: MT7 11-4	☐ Homework Help Online Keyword: MT7 11-4	☐ Homework Help Online Keyword: MT7 11-4	☐ Homework Help Online Keyword: MT7 11-4
☐ *Lesson Tutorial Video* 11-4	☐ *Lesson Tutorial Video* 11-4	☐ *Lesson Tutorial Video* 11-4	☐ *Lesson Tutorial Video* 11-4
☐ Reading Strategies 11-4 CRB	☐ Problem Solving 11-4 CRB	☐ Problem Solving 11-4 CRB	☐ Reading Strategies 11-4 CRB
☐ *Questioning Strategies* pp. 170–171			
☐ *IDEA Works!* 11-4			☐ *Multilingual Glossary*

ASSESSMENT
☐ Lesson Quiz, TE p. 603 and DT 11-4   ☐ State-Specific Test Prep Online Keyword: MT7 TestPrep

Teacher's Name _____ Class _____ Date _____

Lesson Plan 11-5
Solving Two-Step Inequalities pp. 604–607   Day _____

Objective Students solve two-step inequalities and graph the solutions of an inequality on a number line.

> **NCTM Standards:** Understand meanings of operations and how they relate to one another; Represent and analyze mathematical situations and structures using algebraic symbols.

Pacing
- [ ] 45-minute Classes: 1 day   [ ] 90-minute Classes: 1/2 day   [ ] Other_____

WARM UP
- [ ] Warm Up TE p. 604 and Daily Transparency 11-5
- [ ] Problem of the Day TE p. 604 and Daily Transparency 11-5
- [ ] Countdown to Testing Transparency Week 24

TEACH
- [ ] Lesson Presentation CD-ROM 11-5
- [ ] Alternate Opener, Explorations Transparency 11-5, TE p. 604, and Exploration 11-5
- [ ] Reaching All Learners TE p. 605
- [ ] *Hands-On Lab Activities* 11-5
- [ ] *Know-It Notebook* 11-5

PRACTICE AND APPLY
- [ ] Example 1: Average: 1–6, 14–19, 30–32, 47–54 Advanced: 14–19, 27–30, 45, 47–54
- [ ] Example 2: Average: 1–12, 14–25, 30–35, 47–54 Advanced: 14–25, 33–39, 45–54
- [ ] Example 3: Average: 1–26, 30–35, 40–42, 47–54 Advanced: 14–26, 33–54

REACHING ALL LEARNERS – Differentiated Instruction for students with

Developing Knowledge	On-level Knowledge	Advanced Knowledge	English Language Development
[ ] Critical Thinking TE p. 605	[ ] Critical Thinking TE p. 605	[ ] Critical Thinking TE p. 605	[ ] Critical Thinking TE p. 605
[ ] Practice A 11-5 CRB	[ ] Practice B 11-5 CRB	[ ] Practice C 11-5 CRB	[ ] Practice A, B, or C 11-5 CRB
[ ] Reteach 11-5 CRB	[ ] Puzzles, Twisters & Teasers 11-5 CRB	[ ] Challenge 11-5 CRB	[ ] *Success for ELL* 11-5
[ ] Homework Help Online Keyword: MT7 11-5	[ ] Homework Help Online Keyword: MT7 11-5	[ ] Homework Help Online Keyword: MT7 11-5	[ ] Homework Help Online Keyword: MT7 11-5
[ ] *Lesson Tutorial Video* 11-5	[ ] *Lesson Tutorial Video* 11-5	[ ] *Lesson Tutorial Video* 11-5	[ ] *Lesson Tutorial Video* 11-5
[ ] Reading Strategies 11-5 CRB	[ ] Problem Solving 11-5 CRB	[ ] Problem Solving 11-5 CRB	[ ] Reading Strategies 11-5 CRB
[ ] *Questioning Strategies* pp. 172–173			
[ ] IDEA Works! 11-5			[ ] *Multilingual Glossary*

ASSESSMENT
- [ ] Lesson Quiz, TE p. 607 and DT 11-5   [ ] State-Specific Test Prep Online Keyword: MT7 TestPrep

Holt Mathematics

Teacher's Name _____ Class _____ Date _____

Lesson Plan 11-6
Systems of Equations pp. 608–611   Day _____

Objective Students solve systems of equations.

> **NCTM Standards:** Represent and analyze mathematical situations and structures using algebraic symbols.

Pacing
☐ 45-minute Classes: 1 day   ☐ 90-minute Classes: 1/2 day   ☐ Other_____

WARM UP
☐ Warm Up TE p. 608 and Daily Transparency 11-6
☐ Problem of the Day TE p. 608 and Daily Transparency 11-6
☐ Countdown to Testing Transparency Week 25

TEACH
☐ Lesson Presentation CD-ROM 11-6
☐ Alternate Opener, Explorations Transparency 11-6, TE p. 608, and Exploration 11-6
☐ Reaching All Learners TE p. 609
☐ *Know-It Notebook* 11-6

PRACTICE AND APPLY
☐ Example 1: Average: 1–6, 13–18, 26, 27, 41–49 Advanced: 13–18, 26–28, 38, 41–49
☐ Example 2: Average: 1–30, 35, 41–49 Advanced: 13–25, 29–49

REACHING ALL LEARNERS – Differentiated Instruction for students with

Developing Knowledge	On-level Knowledge	Advanced Knowledge	English Language Development
☐ Critical Thinking TE p. 609	☐ Critical Thinking TE p. 609	☐ Critical Thinking TE p. 609	☐ Critical Thinking TE p. 609
☐ Practice A 11-6 CRB	☐ Practice B 11-6 CRB	☐ Practice C 11-6 CRB	☐ Practice A, B, or C 11-6 CRB
☐ Reteach 11-6 CRB	☐ Puzzles, Twisters & Teasers 11-6 CRB	☐ Challenge 11-6 CRB	☐ *Success for ELL* 11-6
☐ Homework Help Online Keyword: MT7 11-6	☐ Homework Help Online Keyword: MT7 11-6	☐ Homework Help Online Keyword: MT7 11-6	☐ Homework Help Online Keyword: MT7 11-6
☐ *Lesson Tutorial Video* 11-6	☐ *Lesson Tutorial Video* 11-6	☐ *Lesson Tutorial Video* 11-6	☐ *Lesson Tutorial Video* 11-6
☐ Reading Strategies 11-6 CRB	☐ Problem Solving 11-6 CRB	☐ Problem Solving 11-6 CRB	☐ Reading Strategies 11-6 CRB
☐ *Questioning Strategies* pp. 174–175			☐ Lesson Vocabulary SE p. 608
☐ *IDEA Works!* 11-6			☐ *Multilingual Glossary*

ASSESSMENT
☐ Lesson Quiz, TE p. 611 and DT 11-6   ☐ State-Specific Test Prep Online Keyword: MT7 TestPrep

Teacher's Name _____ Class _____ Date _____

Lesson Plan 12-1
Graphing Linear Equations pp. 628–632 Day _____

Objective Students identify and graph linear equations.

> **NCTM Standards:** Understand patterns, relations, and functions; Create and use representations to organize, record, and communicate mathematical ideas.

Pacing
☐ 45-minute Classes: 1 day ☐ 90-minute Classes: 1/2 day ☐ Other _____

WARM UP
☐ Warm Up TE p. 628 and Daily Transparency 12-1
☐ Problem of the Day TE p. 628 and Daily Transparency 12-1

TEACH
☐ Lesson Presentation CD-ROM 12-1
☐ Alternate Opener, Explorations Transparency 12-1, TE p. 628, and Exploration 12-1
☐ Reaching All Learners TE p. 629
☐ *Hands-On Lab Activities* 12-1
☐ *Technology Lab Activities* 12-1
☐ *Know-It Notebook* 12-1

PRACTICE AND APPLY
☐ Example 1: Average: 1–3, 5–10, 15–20, 30–37 Advanced: 5–10, 18–23, 28–37
☐ Example 2: Average: 1–13, 15–20, 30–37 Advanced: 5–14, 20–37

REACHING ALL LEARNERS – Differentiated Instruction for students with

Developing Knowledge	On-level Knowledge	Advanced Knowledge	English Language Development
☐ Critical Thinking TE p. 629	☐ Critical Thinking TE p. 629	☐ Critical Thinking TE p. 629	☐ Critical Thinking TE p. 629
☐ Practice A 12-1 CRB	☐ Practice B 12-1 CRB	☐ Practice C 12-1 CRB	☐ Practice A, B, or C 12-1 CRB
☐ Reteach 12-1 CRB	☐ Puzzles, Twisters & Teasers 12-1 CRB	☐ Challenge 12-1 CRB	☐ *Success for ELL* 12-1
☐ Homework Help Online Keyword: MT7 12-1	☐ Homework Help Online Keyword: MT7 12-1	☐ Homework Help Online Keyword: MT7 12-1	☐ Homework Help Online Keyword: MT7 12-1
☐ *Lesson Tutorial Video* 12-1	☐ *Lesson Tutorial Video* 12-1	☐ *Lesson Tutorial Video* 12-1	☐ *Lesson Tutorial Video* 12-1
☐ Reading Strategies 12-1 CRB	☐ Problem Solving 12-1 CRB	☐ Problem Solving 12-1 CRB	☐ Reading Strategies 12-1 CRB
☐ *Questioning Strategies* pp. 176–177	☐ Number Sense TE p. 629	☐ Number Sense TE p. 629	☐ Lesson Vocabulary SE p. 628
☐ *IDEA Works!* 12-1			☐ *Multilingual Glossary*

ASSESSMENT
☐ Lesson Quiz, TE p. 632 and DT 12-1 ☐ State-Specific Test Prep Online Keyword: MT7 TestPrep

Teacher's Name _____ Class _____ Date _____

Lesson Plan 12-2
Slope of a Line pp. 633–637   Day _____

Objective Students find the slope of a line and use slope to understand and draw graphs.

> **NCTM Standards:** Understand patterns, relations, and functions; Analyze change in various contexts; Specify locations and describe spatial relationships using coordinate geometry and other representational systems.

Pacing
☐ 45-minute Classes: 1 day   ☐ 90-minute Classes: 1/2 day   ☐ Other_____

WARM UP
☐ Warm Up TE p. 633 and Daily Transparency 12-2
☐ Problem of the Day TE p. 633 and Daily Transparency 12-2

TEACH
☐ Lesson Presentation CD-ROM 12-2
☐ Alternate Opener, Explorations Transparency 12-2, TE p. 633, and Exploration 12-2
☐ Reaching All Learners TE p. 634
☐ Teaching Transparency 12-2
☐ *Technology Lab Activities* 12-2
☐ *Know-It Notebook* 12-2

PRACTICE AND APPLY
☐ Example 1: Average: 1–3, 8–13, 30–36 Advanced: 8–13, 27, 30–36
☐ Example 2: Average: 1–6, 8–16, 30–36 Advanced: 8–16, 27–36
☐ Example 3: Average: 1–23, 25, 30–36 Advanced: 8–36

REACHING ALL LEARNERS – Differentiated Instruction for students with

Developing Knowledge	On-level Knowledge	Advanced Knowledge	English Language Development
☐ Inclusion TE p. 634	☐ Multiple Representations TE p. 634	☐ Multiple Representations TE p. 634	☐ Multiple Representations TE p. 634
☐ Practice A 12-2 CRB	☐ Practice B 12-2 CRB	☐ Practice C 12-2 CRB	☐ Practice A, B, or C 12-2 CRB
☐ Reteach 12-2 CRB	☐ Puzzles, Twisters & Teasers 12-2 CRB	☐ Challenge 12-2 CRB	☐ *Success for ELL* 12-2
☐ Homework Help Online Keyword: MT7 12-2	☐ Homework Help Online Keyword: MT7 12-2	☐ Homework Help Online Keyword: MT7 12-2	☐ Homework Help Online Keyword: MT7 12-2
☐ *Lesson Tutorial Video* 12-2	☐ *Lesson Tutorial Video* 12-2	☐ *Lesson Tutorial Video* 12-2	☐ *Lesson Tutorial Video* 12-2
☐ Reading Strategies 12-2 CRB	☐ Problem Solving 12-2 CRB	☐ Problem Solving 12-2 CRB	☐ Reading Strategies 12-2 CRB
☐ *Questioning Strategies* pp. 178–179			
☐ *IDEA Works!* 12-2			☐ *Multilingual Glossary*

ASSESSMENT
☐ Lesson Quiz, TE p. 637 and DT 12-2   ☐ State-Specific Test Prep Online Keyword: MT7 TestPrep

Teacher's Name _____ Class _____ Date _____

Lesson Plan 12-3
Using Slopes and Intercepts pp. 638–642 Day _____

Objective Students use slopes and intercepts to graph linear equations.

> **NCTM Standards:** Represent and analyze mathematical situations and structures using algebraic symbols; Specify locations and describe spatial relationships using coordinate geometry and other representational systems.

Pacing
☐ 45-minute Classes: 1 day ☐ 90-minute Classes: 1/2 day ☐ Other_____

WARM UP
☐ Warm Up TE p. 638 and Daily Transparency 12-3
☐ Problem of the Day TE p. 638 and Daily Transparency 12-3

TEACH
☐ Lesson Presentation CD-ROM 12-3
☐ Alternate Opener, Explorations Transparency 12-3, TE p. 638, and Exploration 12-3
☐ Reaching All Learners TE p. 639
☐ Teaching Transparency 12-3
☐ *Hands-On Lab Activities* 12-3
☐ *Know-It Notebook* 12-3

PRACTICE AND APPLY
☐ Example 1: Average: 1–4, 13–16, 28, 33–41 Advanced: 13–16, 25–28, 33–41
☐ Example 2: Average: 1–8, 13–20, 28, 33–41 Advanced: 13–20, 26–29, 33–41
☐ Example 3: Average: 1–9, 13–21, 28, 31, 33–41 Advanced: 13–21, 27–31, 33–41
☐ Example 4: Average: 1–24, 28, 31, 33–41 Advanced: 13–24, 27–41

REACHING ALL LEARNERS – Differentiated Instruction for students with

Developing Knowledge	On-level Knowledge	Advanced Knowledge	English Language Development
☐ Modeling TE p. 639	☐ Modeling TE p. 639	☐ Modeling TE p. 639	☐ Modeling TE p. 639
☐ Practice A 12-3 CRB	☐ Practice B 12-3 CRB	☐ Practice C 12-3 CRB	☐ Practice A, B, or C 12-3 CRB
☐ Reteach 12-3 CRB	☐ Puzzles, Twisters & Teasers 12-3 CRB	☐ Challenge 12-3 CRB	☐ *Success for ELL* 12-3
☐ Homework Help Online Keyword: MT7 12-3	☐ Homework Help Online Keyword: MT7 12-3	☐ Homework Help Online Keyword: MT7 12-3	☐ Homework Help Online Keyword: MT7 12-3
☐ *Lesson Tutorial Video* 12-3	☐ *Lesson Tutorial Video* 12-3	☐ *Lesson Tutorial Video* 12-3	☐ *Lesson Tutorial Video* 12-3
☐ Reading Strategies 12-3 CRB	☐ Problem Solving 12-3 CRB	☐ Problem Solving 12-3 CRB	☐ Reading Strategies 12-3 CRB
☐ *Questioning Strategies* pp. 180–181			☐ Lesson Vocabulary SE p. 638
☐ *IDEA Works!* 12-3			☐ Multilingual Glossary

ASSESSMENT
☐ Lesson Quiz, TE p. 642 and DT 12-3 ☐ State-Specific Test Prep Online Keyword: MT7 TestPrep

Teacher's Name _____ Class _____ Date _____

Lesson Plan 12-4
Point-Slope Form pp. 644–647   Day _____

Objective Students find the equation of a line given one point and the slope.

> **NCTM Standards:** Represent and analyze mathematical situations and structures using algebraic symbols; Create and use representations to organize, record, and communicate mathematical ideas.

Pacing
☐ 45-minute Classes: 1 day   ☐ 90-minute Classes: 1/2 day   ☐ Other_____

WARM UP
☐ Warm Up TE p. 644 and Daily Transparency 12-4
☐ Problem of the Day TE p. 70 and Daily Transparency 12-4

TEACH
☐ Lesson Presentation CD-ROM 12-4
☐ Alternate Opener, Explorations Transparency 12-4, TE p. 70, and Exploration 12-4
☐ Reaching All Learners TE p. 71
☐ Teaching Transparency 12-4
☐ *Know-It Notebook* 12-4

PRACTICE AND APPLY
☐ Example 1: Average: 1–6, 10–15, 29–36 Advanced: 10–15, 29–36
☐ Example 2: Average: 1–8, 10–17, 20, 29–36 Advanced: 10–17, 19–23, 27–36
☐ Example 3: Average: 1–18, 20, 24, 25, 29–36 Advanced: 10–36

REACHING ALL LEARNERS – Differentiated Instruction for students with

Developing Knowledge	On-level Knowledge	Advanced Knowledge	English Language Development
☐ Multiple Representations TE p. 645	☐ Multiple Representations TE p. 645	☐ Multiple Representations TE p. 645	☐ Multiple Representations TE p. 645
☐ Practice A 12-4 CRB	☐ Practice B 12-4 CRB	☐ Practice C 12-4 CRB	☐ Practice A, B, or C 12-4 CRB
☐ Reteach 12-4 CRB	☐ Puzzles, Twisters & Teasers 12-4 CRB	☐ Challenge 12-4 CRB	☐ *Success for ELL* 12-4
☐ Homework Help Online Keyword: MT7 12-4	☐ Homework Help Online Keyword: MT7 12-4	☐ Homework Help Online Keyword: MT7 12-4	☐ Homework Help Online Keyword: MT7 12-4
☐ *Lesson Tutorial Video* 12-4	☐ *Lesson Tutorial Video* 12-4	☐ *Lesson Tutorial Video* 12-4	☐ *Lesson Tutorial Video* 12-4
☐ Reading Strategies 12-4 CRB	☐ Problem Solving 12-4 CRB	☐ Problem Solving 12-4 CRB	☐ Reading Strategies 12-4 CRB
☐ *Questioning Strategies* pp. 182–183			☐ Lesson Vocabulary SE p. 644
☐ *IDEA Works!* 12-4			☐ *Multilingual Glossary*

ASSESSMENT
☐ Lesson Quiz, TE p. 647 and DT 12-4   ☐ State-Specific Test Prep Online Keyword: MT7 TestPrep

Teacher's Name _____ Class _____ Date _____

Lesson Plan 12-5
Direct Variation pp. 650–654   Day _____

Objective Students recognize direct variation by graphing tables of data and checking for constant ratios.

> **NCTM Standards:** Understand patterns, relations, and functions; Use representations to model and interpret physical, social, and mathematical phenomena.

Pacing
☐ 45-minute Classes: 1 day   ☐ 90-minute Classes: 1/2 day   ☐ Other_____

WARM UP
☐ Warm Up TE p. 650 and Daily Transparency 12-5
☐ Problem of the Day TE p. 650 and Daily Transparency 12-5

TEACH
☐ Lesson Presentation CD-ROM 12-5
☐ Alternate Opener, Explorations Transparency 12-5, TE p. 650, and Exploration 12-5
☐ Reaching All Learners TE p. 651
☐ Teaching Transparency 12-5
☐ *Hands-On Lab Activities* 12-5
☐ *Know-It Notebook* 12-5

PRACTICE AND APPLY
☐ Example 1: Average: 1, 9, 17–20, 26–32 Advanced: 9, 17–21, 24, 26–32
☐ Example 2: Average: 1–7, 9–15, 17–20, 26–32 Advanced: 9–15, 17–21, 24, 26–32
☐ Example 3: Average: 1–20, 26–32 Advanced: 9–32

REACHING ALL LEARNERS – Differentiated Instruction for students with

Developing Knowledge	On-level Knowledge	Advanced Knowledge	English Language Development
☐ Kinesthetic Experience TE p. 651	☐ Kinesthetic Experience TE p. 651	☐ Kinesthetic Experience TE p. 651	☐ Kinesthetic Experience TE p. 651
☐ Practice A 12-5 CRB	☐ Practice B 12-5 CRB	☐ Practice C 12-5 CRB	☐ Practice A, B, or C 12-5 CRB
☐ Reteach 12-5 CRB	☐ Puzzles, Twisters & Teasers 12-5 CRB	☐ Challenge 12-5 CRB	☐ *Success for ELL* 12-5
☐ Homework Help Online Keyword: MT7 12-5	☐ Homework Help Online Keyword: MT7 12-5	☐ Homework Help Online Keyword: MT7 12-5	☐ Homework Help Online Keyword: MT7 12-5
☐ *Lesson Tutorial Video* 12-5	☐ *Lesson Tutorial Video* 12-5	☐ *Lesson Tutorial Video* 12-5	☐ *Lesson Tutorial Video* 12-5
☐ Reading Strategies 12-5 CRB	☐ Problem Solving 12-5 CRB	☐ Problem Solving 12-5 CRB	☐ Reading Strategies 12-5 CRB
☐ *Questioning Strategies* pp. 184–185			☐ Lesson Vocabulary SE p. 650
☐ *IDEA Works!* 12-5			☐ *Multilingual Glossary*

ASSESSMENT
☐ Lesson Quiz, TE p. 654 and DT 12-5   ☐ State-Specific Test Prep Online Keyword: MT7 TestPrep

Holt Mathematics

Teacher's Name _____ Class _____ Date _____

Lesson Plan 12-6
Graphing Inequalities in Two Variables pp. 655–659   Day _____

Objective Students graph inequalities on the coordinate plane.

> **NCTM Standards:** Use mathematical models to represent and understand quantitative relationships; Use representations to model and interpret physical, social, and mathematical phenomena.

Pacing
☐ 45-minute Classes: 1 day   ☐ 90-minute Classes: 1/2 day   ☐ Other_____

WARM UP
☐ Warm Up TE p. 655 and Daily Transparency 12-6
☐ Problem of the Day TE p. 655 and Daily Transparency 12-6

TEACH
☐ Lesson Presentation CD-ROM 12-6
☐ Alternate Opener, Explorations Transparency 12-6, TE p. 655, and Exploration 12-6
☐ Reaching All Learners TE p. 656
☐ Teaching Transparency 12-6
☐ *Hands-On Lab Activities* 12-6
☐ *Technology Lab Activities* 12-6
☐ *Know-It Notebook* 12-6

PRACTICE AND APPLY
☐ Example 1: Average: 1–6, 8–13, 15, 18–23, 28–34 Advanced: 8–13, 15, 18–23, 25–34
☐ Example 2: Average: 1–23, 28–34 Advanced: 8–34

REACHING ALL LEARNERS – Differentiated Instruction for students with

Developing Knowledge	On-level Knowledge	Advanced Knowledge	English Language Development
☐ Cognitive Strategies TE p. 656	☐ Cognitive Strategies TE p. 656	☐ Cognitive Strategies TE p. 656	☐ Cognitive Strategies TE p. 656
☐ Practice A 12-6 CRB	☐ Practice B 12-6 CRB	☐ Practice C 12-6 CRB	☐ Practice A, B, or C 12-6 CRB
☐ Reteach 12-6 CRB	☐ Puzzles, Twisters & Teasers 12-6 CRB	☐ Challenge 12-6 CRB	☐ *Success for ELL* 12-6
☐ Homework Help Online Keyword: MT7 12-6	☐ Homework Help Online Keyword: MT7 12-6	☐ Homework Help Online Keyword: MT7 12-6	☐ Homework Help Online Keyword: MT7 12-6
☐ *Lesson Tutorial Video* 12-6	☐ *Lesson Tutorial Video* 12-6	☐ *Lesson Tutorial Video* 12-6	☐ *Lesson Tutorial Video* 12-6
☐ Reading Strategies 12-6 CRB	☐ Problem Solving 12-6 CRB	☐ Problem Solving 12-6 CRB	☐ Reading Strategies 12-6 CRB
☐ *Questioning Strategies* pp. 186–187			☐ Lesson Vocabulary SE p. 655
☐ *IDEA Works!* 12-6			☐ *Multilingual Glossary*

ASSESSMENT
☐ Lesson Quiz, TE p. 659 and DT 12-6   ☐ State-Specific Test Prep Online Keyword: MT7 TestPrep

Teacher's Name _____ Class _____ Date _____

Lesson Plan 12-7
Lines of Best Fit pp. 660–663 Day _____

Objective Students recognize relationships in data and find the equation of a line of best fit.

> **NCTM Standards:** Understand patterns, relations, and functions; Develop and evaluate inferences and predictions that are based on data; Use representations to model and interpret physical, social, and mathematical phenomena.

Pacing
☐ 45-minute Classes: 1 day   ☐ 90-minute Classes: 1/2 day   ☐ Other _____

WARM UP
☐ Warm Up TE p. 660 and Daily Transparency 12-7
☐ Problem of the Day TE p. 660 and Daily Transparency 12-7

TEACH
☐ Lesson Presentation CD-ROM 12-7
☐ Alternate Opener, Explorations Transparency 12-7, TE p. 660, and Exploration 12-7
☐ Reaching All Learners TE p. 661
☐ *Hands-On Lab Activities* 12-7
☐ *Know-It Notebook* 12-7

PRACTICE AND APPLY
☐ Example 1: Average: 1, 2, 4, 5, 7–10, 15–24 Advanced: 4, 5, 7–10, 14–24
☐ Example 2: Average: 1–10, 15–24 Advanced: 4–24

REACHING ALL LEARNERS – Differentiated Instruction for students with

Developing Knowledge	On-level Knowledge	Advanced Knowledge	English Language Development
☐ Cooperative Learning TE p. 661	☐ Cooperative Learning TE p. 661	☐ Cooperative Learning TE p. 661	☐ Cooperative Learning TE p. 661
☐ Practice A 12-7 CRB	☐ Practice B 12-7 CRB	☐ Practice C 12-7 CRB	☐ Practice A, B, or C 12-7 CRB
☐ Reteach 12-7 CRB	☐ Puzzles, Twisters & Teasers 12-7 CRB	☐ Challenge 12-7 CRB	☐ *Success for ELL* 12-7
☐ Homework Help Online Keyword: MT7 12-7	☐ Homework Help Online Keyword: MT7 12-7	☐ Homework Help Online Keyword: MT7 12-7	☐ Homework Help Online Keyword: MT7 12-7
☐ *Lesson Tutorial Video* 12-7	☐ *Lesson Tutorial Video* 12-7	☐ *Lesson Tutorial Video* 12-7	☐ *Lesson Tutorial Video* 12-7
☐ Reading Strategies 12-7 CRB	☐ Problem Solving 12-7 CRB	☐ Problem Solving 12-7 CRB	☐ Reading Strategies 12-7 CRB
☐ *Questioning Strategies* pp. 188–189	☐ Communicating Math TE p. 661	☐ Communicating Math TE p. 661	
☐ *IDEA Works!* 12-7			☐ *Multilingual Glossary*

ASSESSMENT
☐ Lesson Quiz, TE p. 663 and DT 12-7   ☐ State-Specific Test Prep Online Keyword: MT7 TestPrep

Holt Mathematics

Teacher's Name _____ Class _____ Date _____

Lesson Plan 13-1
Terms of Arithmetic Sequences pp. 682–686   Day _____

Objective Students find terms in an arithmetic sequence.

> **NCTM Standards:** Understand patterns, relations, and functions.

Pacing
☐ 45-minute Classes: 1 day   ☐ 90-minute Classes: 1/2 day   ☐ Other_____

WARM UP
☐ Warm Up TE p. 682 and Daily Transparency 13-1
☐ Problem of the Day TE p. 682 and Daily Transparency 13-1

TEACH
☐ Lesson Presentation CD-ROM 13-1
☐ Alternate Opener, Explorations Transparency 13-1, TE p. 682, and Exploration 13-1
☐ Reaching All Learners TE p. 683
☐ Teaching Transparency 13-1
☐ *Know-It Notebook* 13-1

PRACTICE AND APPLY
☐ Example 1: Average: 1–6, 12–17, 23–25, 37–44 Advanced: 12–17, 23–25, 35, 37–44
☐ Example 2: Average: 1–10, 12–21, 23–25, 27–29, 37–44 Advanced: 12–21, 23–30, 34–44
☐ Example 3: Average: 1–25, 27–29, 31, 32, 37–44 Advanced: 12–44

REACHING ALL LEARNERS – Differentiated Instruction for students with

Developing Knowledge	On-level Knowledge	Advanced Knowledge	English Language Development
☐ Critical Thinking TE p. 683	☐ Critical Thinking TE p. 683	☐ Critical Thinking TE p. 683	☐ Critical Thinking TE p. 683
☐ Practice A 13-1 CRB	☐ Practice B 13-1 CRB	☐ Practice C 13-1 CRB	☐ Practice A, B, or C 13-1 CRB
☐ Reteach 13-1 CRB	☐ Puzzles, Twisters & Teasers 13-1 CRB	☐ Challenge 13-1 CRB	☐ *Success for ELL* 13-1
☐ Homework Help Online Keyword: MT7 13-1	☐ Homework Help Online Keyword: MT7 13-1	☐ Homework Help Online Keyword: MT7 13-1	☐ Homework Help Online Keyword: MT7 13-1
☐ *Lesson Tutorial Video* 13-1	☐ *Lesson Tutorial Video* 13-1	☐ *Lesson Tutorial Video* 13-1	☐ *Lesson Tutorial Video* 13-1
☐ Reading Strategies 13-1 CRB	☐ Problem Solving 13-1 CRB	☐ Problem Solving 13-1 CRB	☐ Reading Strategies 13-1 CRB
☐ *Questioning Strategies* pp. 190–191			
☐ *IDEA Works!* 13-1			☐ *Multilingual Glossary*

ASSESSMENT
☐ Lesson Quiz, TE p. 686 and DT 13-1   ☐ State-Specific Test Prep Online Keyword: MT7 TestPrep

Teacher's Name _____ Class _____ Date _____

Lesson Plan 13-2
Terms of Geometric Sequences pp. 687–691   Day _____

Objective Students find terms in a geometric sequence.

> **NCTM Standards:** Understand patterns, relations, and functions.

Pacing
☐ 45-minute Classes: 1 day   ☐ 90-minute Classes: 1/2 day   ☐ Other_____

WARM UP
☐ Warm Up TE p. 687 and Daily Transparency 13-2
☐ Problem of the Day TE p. 687 and Daily Transparency 13-2

TEACH
☐ Lesson Presentation CD-ROM 13-2
☐ Alternate Opener, Explorations Transparency 13-2, TE p. 687, and Exploration 13-2
☐ Reaching All Learners TE p. 688
☐ Teaching Transparency 13-2
☐ *Hands-On Lab Activities* 13-2
☐ *Technology Lab Activities* 13-2
☐ *Know-It Notebook* 13-2

PRACTICE AND APPLY
☐ Example 1: Average: 1–6, 12–17, 25–28, 48–55 Advanced: 12–17, 25–28, 48–55
☐ Example 2: Average: 1–10, 12–23, 25–34, 48–55 Advanced: 1–23, 25–39, 45–55
☐ Example 3: Average: 1–34, 40–42, 48–55 Advanced: 12–28, 32–55

REACHING ALL LEARNERS – Differentiated Instruction for students with

Developing Knowledge	On-level Knowledge	Advanced Knowledge	English Language Development
☐ Critical Thinking TE p. 688	☐ Critical Thinking TE p. 688	☐ Critical Thinking TE p. 688	☐ Critical Thinking TE p. 688
☐ Practice A 13-2 CRB	☐ Practice B 13-2 CRB	☐ Practice C 13-2 CRB	☐ Practice A, B, or C 13-2 CRB
☐ Reteach 13-2 CRB	☐ Puzzles, Twisters & Teasers 13-2 CRB	☐ Challenge 13-2 CRB	☐ *Success for ELL* 13-2
☐ Homework Help Online Keyword: MT7 13-2	☐ Homework Help Online Keyword: MT7 13-2	☐ Homework Help Online Keyword: MT7 13-2	☐ Homework Help Online Keyword: MT7 13-2
☐ *Lesson Tutorial Video* 13-2	☐ *Lesson Tutorial Video* 13-2	☐ *Lesson Tutorial Video* 13-2	☐ *Lesson Tutorial Video* 13-2
☐ Reading Strategies 13-2 CRB	☐ Problem Solving 13-2 CRB	☐ Problem Solving 13-2 CRB	☐ Reading Strategies 13-2 CRB
☐ *Questioning Strategies* pp. 192–193			☐ Lesson Vocabulary SE p. 687
☐ *IDEA Works!* 13-2			☐ *Multilingual Glossary*

ASSESSMENT
☐ Lesson Quiz, TE p. 691 and DT 13-2   ☐ State-Specific Test Prep Online Keyword: MT7 TestPrep

Teacher's Name _____ Class _____ Date _____

Lesson Plan 13-3
Other Sequences pp. 693–697   Day _____

Objective Students find patterns in sequences.

> **NCTM Standards:** Understand patterns, relations, and functions.

Pacing
☐ 45-minute Classes: 1 day   ☐ 90-minute Classes: 1/2 day   ☐ Other_____

WARM UP
☐ Warm Up TE p. 693 and Daily Transparency 13-3
☐ Problem of the Day TE p. 693 and Daily Transparency 13-3

TEACH
☐ Lesson Presentation CD-ROM 13-3
☐ Alternate Opener, Explorations Transparency 13-3, TE p. 693, and Exploration 13-3
☐ Reaching All Learners TE p. 694
☐ Teaching Transparency 13-3
☐ *Know-It Notebook* 13-3

PRACTICE AND APPLY
☐ Example 1: Average: 1–4, 13–16, 27, 32–39 Advanced: 13–16, 27, 32–39
☐ Example 2: Average: 1–8, 13–20, 27, 32–39 Advanced: 13–20, 27–39
☐ Example 3: Average: 1–11, 13–23, 27, 32–39 Advanced: 13–23, 27–39
☐ Example 4: Average: 1–27, 32–39 Advanced: 13–39

REACHING ALL LEARNERS – Differentiated Instruction for students with

Developing Knowledge	On-level Knowledge	Advanced Knowledge	English Language Development
☐ Concrete Manipulatives TE p. 694	☐ Concrete Manipulatives TE p. 694	☐ Concrete Manipulatives TE p. 694	☐ Concrete Manipulatives TE p. 694
☐ Practice A 13-3 CRB	☐ Practice B 13-3 CRB	☐ Practice C 13-3 CRB	☐ Practice A, B, or C 13-3 CRB
☐ Reteach 13-3 CRB	☐ Puzzles, Twisters & Teasers 13-3 CRB	☐ Challenge 13-3 CRB	☐ *Success for ELL* 13-3
☐ Homework Help Online Keyword: MT7 13-3	☐ Homework Help Online Keyword: MT7 13-3	☐ Homework Help Online Keyword: MT7 13-3	☐ Homework Help Online Keyword: MT7 13-3
☐ *Lesson Tutorial Video* 13-3	☐ *Lesson Tutorial Video* 13-3	☐ *Lesson Tutorial Video* 13-3	☐ *Lesson Tutorial Video* 13-3
☐ Reading Strategies 13-3 CRB	☐ Problem Solving 13-3 CRB	☐ Problem Solving 13-3 CRB	☐ Reading Strategies 13-3 CRB
☐ *Questioning Strategies* pp. 194–195			☐ Lesson Vocabulary SE p. 693
☐ *IDEA Works!* 13-3			☐ *Multilingual Glossary*

ASSESSMENT
☐ Lesson Quiz, TE p. 697 and DT 13-3   ☐ State-Specific Test Prep Online Keyword: MT7 TestPrep

Teacher's Name _____ Class _____ Date _____

Lesson Plan 13-4
Linear Functions pp. 700–703   Day _____

Objective Students identify linear functions.

> **NCTM Standards:** Understand patterns, relations, and functions; Understand how mathematical ideas interconnect and build on one another to produce a coherent whole.

Pacing
☐ 45-minute Classes: 1 day   ☐ 90-minute Classes: 1/2 day   ☐ Other_____

WARM UP
☐ Warm Up TE p. 700 and Daily Transparency 13-4
☐ Problem of the Day TE p. 700 and Daily Transparency 13-4

TEACH
☐ Lesson Presentation CD-ROM 13-4
☐ Alternate Opener, Explorations Transparency 13-4, TE p. 700, and Exploration 13-4
☐ Reaching All Learners TE p. 701
☐ Teaching Transparency 13-4
☐ *Know-It Notebook* 13-4

PRACTICE AND APPLY
☐ Example 1: Average: 1–3, 7–9, 20–29 Advanced: 7–9, 17–29
☐ Example 2: Average: 1–5, 7–11, 20–29 Advanced: 7–11, 17–29
☐ Example 3: Average: 1–12, 14, 15, 20–29 Advanced: 7–29

REACHING ALL LEARNERS – Differentiated Instruction for students with

Developing Knowledge	On-level Knowledge	Advanced Knowledge	English Language Development
☐ Modeling TE p. 701	☐ Modeling TE p. 701	☐ Modeling TE p. 701	☐ Modeling TE p. 701
☐ Practice A 13-4 CRB	☐ Practice B 13-4 CRB	☐ Practice C 13-4 CRB	☐ Practice A, B, or C 13-4 CRB
☐ Reteach 13-4 CRB	☐ Puzzles, Twisters & Teasers 13-4 CRB	☐ Challenge 13-4 CRB	☐ *Success for ELL* 13-4
☐ Homework Help Online Keyword: MT7 13-4	☐ Homework Help Online Keyword: MT7 13-4	☐ Homework Help Online Keyword: MT7 13-4	☐ Homework Help Online Keyword: MT7 13-4
☐ *Lesson Tutorial Video* 13-4	☐ *Lesson Tutorial Video* 13-4	☐ *Lesson Tutorial Video* 13-4	☐ *Lesson Tutorial Video* 13-4
☐ Reading Strategies 13-4 CRB	☐ Problem Solving 13-4 CRB	☐ Problem Solving 13-4 CRB	☐ Reading Strategies 13-4 CRB
☐ *Questioning Strategies* pp. 196–197	☐ Multiple Representations TE p. 701	☐ Multiple Representations TE p. 701	☐ Lesson Vocabulary SE p. 700
☐ *IDEA Works!* 13-4			☐ *Multilingual Glossary*

ASSESSMENT
☐ Lesson Quiz, TE p. 703 and DT 13-4   ☐ State-Specific Test Prep Online Keyword: MT7 TestPrep

Teacher's Name _____ Class _____ Date _____

Lesson Plan 13-5
Exponential Functions pp. 704–707 Day _____

Objective Students identify and graph exponential functions.

> **NCTM Standards:** Understand patterns, relations, and functions; Understand how mathematical ideas interconnect and build on one another to produce a coherent whole.

Pacing
☐ 45-minute Classes: 1 day   ☐ 90-minute Classes: 1/2 day   ☐ Other_____

WARM UP
☐ Warm Up TE p. 704 and Daily Transparency 13-5
☐ Problem of the Day TE p. 704 and Daily Transparency 13-5

TEACH
☐ Lesson Presentation CD-ROM 13-5
☐ Alternate Opener, Explorations Transparency 13-5, TE p. 704, and Exploration 13-5
☐ Reaching All Learners TE p. 705
☐ *Technology Lab Activities* 13-5
☐ *Know-It Notebook* 13-5

PRACTICE AND APPLY
☐ Example 1: Average: 1–6, 9–14, 17–20, 24, 33–38 Advanced: 9–14, 17–26, 33–38
☐ Example 2: Average: 1–7, 9–15, 17–20, 24, 33–38 Advanced: 9–15, 17–26, 33–38
☐ Example 3: Average: 1–20, 24, 28–30, 33–38 Advanced: 9–38

REACHING ALL LEARNERS – Differentiated Instruction for students with

Developing Knowledge	On-level Knowledge	Advanced Knowledge	English Language Development
☐ Cognitive Strategies TE p. 705	☐ Cognitive Strategies TE p. 705	☐ Cognitive Strategies TE p. 705	☐ Cognitive Strategies TE p. 705
☐ Practice A 13-5 CRB	☐ Practice B 13-5 CRB	☐ Practice C 13-5 CRB	☐ Practice A, B, or C 13-5 CRB
☐ Reteach 13-5 CRB	☐ Puzzles, Twisters & Teasers 13-5 CRB	☐ Challenge 13-5 CRB	☐ *Success for ELL* 13-5
☐ Homework Help Online Keyword: MT7 13-5	☐ Homework Help Online Keyword: MT7 13-5	☐ Homework Help Online Keyword: MT7 13-5	☐ Homework Help Online Keyword: MT7 13-5
☐ *Lesson Tutorial Video* 13-5	☐ *Lesson Tutorial Video* 13-5	☐ *Lesson Tutorial Video* 13-5	☐ *Lesson Tutorial Video* 13-5
☐ Reading Strategies 13-5 CRB	☐ Problem Solving 13-5 CRB	☐ Problem Solving 13-5 CRB	☐ Reading Strategies 13-5 CRB
☐ *Questioning Strategies* pp. 198–199			☐ Lesson Vocabulary SE p. 704
☐ *IDEA Works!* 13-5			☐ *Multilingual Glossary*

ASSESSMENT
☐ Lesson Quiz, TE p. 707 and DT 13-5   ☐ State-Specific Test Prep Online Keyword: MT7 TestPrep

Teacher's Name _____ Class _____ Date _____

Lesson Plan 13-6
Quadratic Functions pp. 708–711   Day _____

Objective Students identify and graph quadratic functions.

> **NCTM Standards:** Understand patterns, relations, and functions; Understand how mathematical ideas interconnect and build on one another to produce a coherent whole.

Pacing
☐ 45-minute Classes: 1 day   ☐ 90-minute Classes: 1/2 day   ☐ Other_____

WARM UP
☐ Warm Up TE p. 708 and Daily Transparency 13-6
☐ Problem of the Day TE p. 708 and Daily Transparency 13-6

TEACH
☐ Lesson Presentation CD-ROM 13-6
☐ Alternate Opener, Explorations Transparency 13-6, TE p. 708, and Exploration 13-6
☐ Reaching All Learners TE p. 709
☐ *Hands-On Activities* 13-6
☐ *Technology Activities* 13-6
☐ *Know-It Notebook* 13-6

PRACTICE AND APPLY
☐ Example 1: Average: 1–3, 5–7, 9–18, 28–35 Advanced: 5–7, 9–19, 22, 26–35
☐ Example 2: Average: 1–18, 20, 21, 28–35 Advanced: 5–12, 17–35

REACHING ALL LEARNERS – Differentiated Instruction for students with

Developing Knowledge	On-level Knowledge	Advanced Knowledge	English Language Development
☐ Visual Cues TE p. 709	☐ Visual Cues TE p. 709	☐ Visual Cues TE p. 709	☐ Visual Cues TE p. 709
☐ Practice A 13-6 CRB	☐ Practice B 13-6 CRB	☐ Practice C 13-6 CRB	☐ Practice A, B, or C 13-6 CRB
☐ Reteach 13-6 CRB	☐ Puzzles, Twisters & Teasers 13-6 CRB	☐ Challenge 13-6 CRB	☐ *Success for ELL* 13-6
☐ Homework Help Online Keyword: MT7 13-6	☐ Homework Help Online Keyword: MT7 13-6	☐ Homework Help Online Keyword: MT7 13-6	☐ Homework Help Online Keyword: MT7 13-6
☐ *Lesson Tutorial Video* 13-6	☐ *Lesson Tutorial Video* 13-6	☐ *Lesson Tutorial Video* 13-6	☐ *Lesson Tutorial Video* 13-6
☐ Reading Strategies 13-6 CRB	☐ Problem Solving 13-6 CRB	☐ Problem Solving 13-6 CRB	☐ Reading Strategies 13-6 CRB
☐ *Questioning Strategies* pp. 200–201			☐ Lesson Vocabulary SE p. 708
☐ *IDEA Works!* 13-6			☐ Multilingual Glossary

ASSESSMENT
☐ Lesson Quiz, TE p. 711 and DT 13-6   ☐ State-Specific Test Prep Online Keyword: MT7 TestPrep

Teacher's Name _____ Class _____ Date _____

Lesson Plan 13-7
Inverse Variation pp. 714–717   Day _____

Objective Students recognize inverse variation by graphing tables of data.

> **NCTM Standards:** Understand patterns, relations, and functions; Use representations to model and interpret physical, social, and mathematical phenomena.

Pacing
☐ 45-minute Classes: 1 day   ☐ 90-minute Classes: 1/2 day   ☐ Other _____

WARM UP
☐ Warm Up TE p. 714 and Daily Transparency 13-7
☐ Problem of the Day TE p. 714 and Daily Transparency 13-7

TEACH
☐ Lesson Presentation CD-ROM 13-7
☐ Alternate Opener, Explorations Transparency 13-7, TE p. 714, and Exploration 13-7
☐ Reaching All Learners TE p. 715
☐ Teaching Transparency 13-7
☐ *Know-It Notebook* 13-7

PRACTICE AND APPLY
☐ Example 1: Average: 1, 2, 8, 9, 25–32 Advanced: 8, 9, 15–17, 23, 25–32
☐ Example 2: Average: 1–6, 8–13, 25–32 Advanced: 8–13, 15–17, 23, 25–32
☐ Example 3: Average: 1–18, 20, 23, 25–32 Advanced: 8–32

REACHING ALL LEARNERS – Differentiated Instruction for students with

Developing Knowledge	On-level Knowledge	Advanced Knowledge	English Language Development
☐ Inclusion TE p. 715	☐ Inclusion TE p. 715	☐ Inclusion TE p. 715	☐ Inclusion TE p. 715
☐ Practice A 13-7 CRB	☐ Practice B 13-7 CRB	☐ Practice C 13-7 CRB	☐ Practice A, B, or C 13-7 CRB
☐ Reteach 13-7 CRB	☐ Puzzles, Twisters & Teasers 13-7 CRB	☐ Challenge 13-7 CRB	☐ *Success for ELL* 13-7
☐ Homework Help Online Keyword: MT7 13-7	☐ Homework Help Online Keyword: MT7 13-7	☐ Homework Help Online Keyword: MT7 13-7	☐ Homework Help Online Keyword: MT7 13-7
☐ *Lesson Tutorial Video* 13-7	☐ *Lesson Tutorial Video* 13-7	☐ *Lesson Tutorial Video* 13-7	☐ *Lesson Tutorial Video* 13-7
☐ Reading Strategies 13-7 CRB	☐ Problem Solving 13-7 CRB	☐ Problem Solving 13-7 CRB	☐ Reading Strategies 13-7 CRB
☐ *Questioning Strategies* pp. 202–203			☐ *Lesson Vocabulary* SE p. 714
☐ *IDEA Works!* 13-7			☐ *Multilingual Glossary*

ASSESSMENT
☐ Lesson Quiz, TE p. 717 and DT 13-7   ☐ State-Specific Test Prep Online Keyword: MT7 TestPrep

Teacher's Name _____ Class _____ Date _____

Lesson Plan 14-1

Polynomials pp. 734–737   Day _____

Objective Students classify polynomials by degree and by the number of terms.

> **NCTM Standards:** Understand and apply basic concepts of probability; Select and use various types of reasoning and methods of proof; Recognize and apply mathematics in contexts outside of mathematics.

Pacing
☐ 45-minute Classes: 1 day   ☐ 90-minute Classes: 1/2 day   ☐ Other_____

WARM UP
☐ Warm Up TE p. 734 and Daily Transparency 14-1
☐ Problem of the Day TE p. 734 and Daily Transparency 14-1

TEACH
☐ Lesson Presentation CD-ROM 14-1
☐ Alternate Opener, Explorations Transparency 14-1, TE p. 734, and Exploration 14-1
☐ Reaching All Learners TE p. 735
☐ *Know-It Notebook* 14-1

PRACTICE AND APPLY
☐ Example 1: Average: 1–4, 13–18, 50–57 Advanced: 13–18, 48, 50–57
☐ Example 2: Average: 1–8, 13–24, 50–57 Advanced: 13–24, 48, 50–57
☐ Example 3: Average: 1–11, 13–30, 33–44, 50–57 Advanced: 13–30, 33–44, 46–48, 50–57
☐ Example 4: Average: 1–45, 50–57 Advanced: 13–57

REACHING ALL LEARNERS – Differentiated Instruction for students with

Developing Knowledge	On-level Knowledge	Advanced Knowledge	English Language Development
☐ Critical Thinking TE p. 735	☐ Critical Thinking TE p. 735	☐ Critical Thinking TE p. 735	☐ Critical Thinking TE p. 735
☐ Practice A 14-1 CRB	☐ Practice B 14-1 CRB	☐ Practice C 14-1 CRB	☐ Practice A, B, or C 14-1 CRB
☐ Reteach 14-1 CRB	☐ Puzzles, Twisters & Teasers 14-1 CRB	☐ Challenge 14-1 CRB	☐ *Success for ELL* 14-1
☐ Homework Help Online Keyword: MT7 14-1	☐ Homework Help Online Keyword: MT7 14-1	☐ Homework Help Online Keyword: MT7 14-1	☐ Homework Help Online Keyword: MT7 14-1
☐ *Lesson Tutorial Video* 14-1	☐ *Lesson Tutorial Video* 14-1	☐ *Lesson Tutorial Video* 14-1	☐ *Lesson Tutorial Video* 14-1
☐ Reading Strategies 14-1 CRB	☐ Problem Solving 14-1 CRB	☐ Problem Solving 14-1 CRB	☐ Reading Strategies 14-1 CRB
☐ *Questioning Strategies* pp. 204–205			☐ Lesson Vocabulary SE p. 734
☐ *IDEA Works!* 14-1			☐ *Multilingual Glossary*

ASSESSMENT
☐ Lesson Quiz, TE p. 737 and DT 14-1   ☐ State-Specific Test Prep Online Keyword: MT7 TestPrep

Teacher's Name _____ Class _____ Date _____

Lesson Plan 14-2
Simplifying Polynomials pp. 740–743 Day _____

Objective Students simplify polynomials.

> **NCTM Standards:** Understand and apply basic concepts of probability; Select and use various types of reasoning and methods of proof.

Pacing
☐ 45-minute Classes: 1 day ☐ 90-minute Classes: 1/2 day ☐ Other _____

WARM UP
☐ Warm Up TE p. 740 and Daily Transparency 14-2
☐ Problem of the Day TE p. 740 and Daily Transparency 14-2

TEACH
☐ Lesson Presentation CD-ROM 14-2
☐ Alternate Opener, Explorations Transparency 14-2, TE p. 740, and Exploration 14-2
☐ Reaching All Learners TE p. 741
☐ Teaching Transparency 14-2
☐ *Know-It Notebook* 14-2

PRACTICE AND APPLY
☐ Example 1: Average: 1, 2, 9, 10, 29–34 Advanced: 9, 10, 29–34
☐ Example 2: Average: 1–4, 9–12, 17, 18, 29–34 Advanced: 9–12, 17, 18, 28–34
☐ Example 3: Average: 1–7, 9–15, 17–22, 29–34 Advanced: 9–15, 19–24, 28–34
☐ Example 4: Average: 1–22, 25, 26, 29–34 Advanced: 9–34

REACHING ALL LEARNERS – Differentiated Instruction for students with

Developing Knowledge	On-level Knowledge	Advanced Knowledge	English Language Development
☐ Kinesthetic Experience TE p. 741	☐ Kinesthetic Experience TE p. 741	☐ Kinesthetic Experience TE p. 741	☐ Kinesthetic Experience TE p. 741
☐ Practice A 14-2 CRB	☐ Practice B 14-2 CRB	☐ Practice C 14-2 CRB	☐ Practice A, B, or C 14-2 CRB
☐ Reteach 14-2 CRB	☐ Puzzles, Twisters & Teasers 14-2 CRB	☐ Challenge 14-2 CRB	☐ *Success for ELL* 14-2
☐ Homework Help Online Keyword: MT7 14-2	☐ Homework Help Online Keyword: MT7 14-2	☐ Homework Help Online Keyword: MT7 14-2	☐ Homework Help Online Keyword: MT7 14-2
☐ *Lesson Tutorial Video* 14-2	☐ *Lesson Tutorial Video* 14-2	☐ *Lesson Tutorial Video* 14-2	☐ *Lesson Tutorial Video* 14-2
☐ Reading Strategies 14-2 CRB	☐ Problem Solving 14-2 CRB	☐ Problem Solving 14-2 CRB	☐ Reading Strategies 14-2 CRB
☐ *Questioning Strategies* pp. 206–207			
☐ *IDEA Works!* 14-2			☐ *Multilingual Glossary*

ASSESSMENT
☐ Lesson Quiz, TE p. 743 and DT 14-2   ☐ State-Specific Test Prep Online Keyword: MT7 TestPrep

Teacher's Name _____ Class _____ Date _____

Lesson Plan 14-3
Adding Polynomials pp. 747–750   Day _____

Objective Students add polynomials.

> **NCTM Standards:** Compute fluently and make reasonable estimates; Develop and evaluate mathematical arguments and proofs; Recognize reasoning and proof as fundamental aspects of mathematics.

Pacing
☐ 45-minute Classes: 1 day   ☐ 90-minute Classes: 1/2 day   ☐ Other_____

WARM UP
☐ Warm Up TE p. 747 and Daily Transparency 14-3
☐ Problem of the Day TE p. 747 and Daily Transparency 14-3

TEACH
☐ Lesson Presentation CD-ROM 14-3
☐ Alternate Opener, Explorations Transparency 14-3, TE p. 747, and Exploration 14-3
☐ Reaching All Learners TE p. 748
☐ *Know-It Notebook* 14-3

PRACTICE AND APPLY
☐ Example 1: Average: 1–3, 8–12, 17, 18, 25, 27–36 Advanced: 8–12, 17, 18, 22, 23, 25–36
☐ Example 2: Average: 1–6, 8–15, 17, 18, 25, 27–36 Advanced: 8–15, 17, 18, 22, 23, 25–36
☐ Example 3: Average: 1–21, 25, 27–36 Advanced: 8–36

REACHING ALL LEARNERS – Differentiated Instruction for students with

Developing Knowledge	On-level Knowledge	Advanced Knowledge	English Language Development
☐ Inclusion TE p. 748	☐ Concrete Manipulatives TE p. 748	☐ Concrete Manipulatives TE p. 748	☐ Concrete Manipulatives TE p. 748
☐ Practice A 14-3 CRB	☐ Practice B 14-3 CRB	☐ Practice C 14-3 CRB	☐ Practice A, B, or C 14-3 CRB
☐ Reteach 14-3 CRB	☐ Puzzles, Twisters & Teasers 14-3 CRB	☐ Challenge 14-3 CRB	☐ *Success for ELL* 14-3
☐ Homework Help Online Keyword: MT7 14-3	☐ Homework Help Online Keyword: MT7 14-3	☐ Homework Help Online Keyword: MT7 14-3	☐ Homework Help Online Keyword: MT7 14-3
☐ *Lesson Tutorial Video* 14-3	☐ *Lesson Tutorial Video* 14-3	☐ *Lesson Tutorial Video* 14-3	☐ *Lesson Tutorial Video* 14-3
☐ Reading Strategies 14-3 CRB	☐ Problem Solving 14-3 CRB	☐ Problem Solving 14-3 CRB	☐ Reading Strategies 14-3 CRB
☐ *Questioning Strategies* pp. 208–209			
☐ *IDEA Works!* 14-3			☐ *Multilingual Glossary*

ASSESSMENT
☐ Lesson Quiz, TE p. 750 and DT 14-3   ☐ State-Specific Test Prep Online Keyword: MT7 TestPrep

Teacher's Name _____ Class _____ Date _____

Lesson Plan 14-4
Subtracting Polynomials pp. 752–755   Day _____

Objective Students subtract polynomials.

> **NCTM Standards:** Understand and apply basic concepts of probability; Select and use various types of reasoning and methods of proof.

Pacing
☐ 45-minute Classes: 1 day   ☐ 90-minute Classes: 1/2 day   ☐ Other_____

WARM UP
☐ Warm Up TE p. 752 and Daily Transparency 14-4
☐ Problem of the Day TE p. 752 and Daily Transparency 14-4

TEACH
☐ Lesson Presentation CD-ROM 14-4
☐ Alternate Opener, Explorations Transparency 14-4, TE p. 752, and Exploration 14-4
☐ Reaching All Learners TE p. 753
☐ *Know-It Notebook* 14-4

PRACTICE AND APPLY
☐ Example 1: Average: 1–6, 14–19, 36–44 Advanced: 14–19, 33, 36–44
☐ Example 2: Average: 1–9, 14–22, 27, 28, 34, 36–44 Advanced: 14–22, 27–29, 33–44
☐ Example 3: Average: 1–12, 14–25, 27–29, 34, 36–44 Advanced: 14–25, 27–29, 33–44
☐ Example 4: Average: 1–29, 32, 34, 36–44 Advanced: 14–44

REACHING ALL LEARNERS – Differentiated Instruction for students with

Developing Knowledge	On-level Knowledge	Advanced Knowledge	English Language Development
☐ Inclusion TE p. 753	☐ Concrete Manipulatives TE p. 753	☐ Concrete Manipulatives TE p. 753	☐ Concrete Manipulatives TE p. 753
☐ Practice A 14-4 CRB	☐ Practice B 14-4 CRB	☐ Practice C 14-4 CRB	☐ Practice A, B, or C 14-4 CRB
☐ Reteach 14-4 CRB	☐ Puzzles, Twisters & Teasers 14-4 CRB	☐ Challenge 14-4 CRB	☐ *Success for ELL* 14-4
☐ Homework Help Online Keyword: MT7 14-4	☐ Homework Help Online Keyword: MT7 14-4	☐ Homework Help Online Keyword: MT7 14-4	☐ Homework Help Online Keyword: MT7 14-4
☐ *Lesson Tutorial Video* 14-4	☐ *Lesson Tutorial Video* 14-4	☐ *Lesson Tutorial Video* 14-4	☐ *Lesson Tutorial Video* 14-4
☐ Reading Strategies 14-4 CRB	☐ Problem Solving 14-4 CRB	☐ Problem Solving 14-4 CRB	☐ Reading Strategies 14-4 CRB
☐ *Questioning Strategies* pp. 210–211			
☐ *IDEA Works!* 14-4			☐ *Multilingual Glossary*

ASSESSMENT
☐ Lesson Quiz, TE p. 755 and DT 14-4   ☐ State-Specific Test Prep Online Keyword: MT7 TestPrep

Teacher's Name _____ Class _____ Date _____

Lesson Plan 14-5
Multiplying Polynomials by Monomials pp. 756–759   Day _____

Objective Students multiply polynomials by monomials.

> **NCTM Standards:** Understand and apply basic concepts of probability.

Pacing
☐ 45-minute Classes: 1 day   ☐ 90-minute Classes: 1/2 day   ☐ Other _____

WARM UP
☐ Warm Up TE p. 756 and Daily Transparency 14-5
☐ Problem of the Day TE p. 756 and Daily Transparency 14-5

TEACH
☐ Lesson Presentation CD-ROM 14-5
☐ Alternate Opener, Explorations Transparency 14-5, TE p. 756, and Exploration 14-5
☐ Reaching All Learners TE p. 757
☐ *Know-It Notebook* 14-5

PRACTICE AND APPLY
☐ Example 1: Average: 1–6, 12–17, 25–27, 42–48 Advanced: 12–17, 25–27, 42–48
☐ Example 2: Average: 1–10, 12–23, 25–34, 42–48 Advanced: 12–23, 25–36, 40–48
☐ Example 3: Average: 1–34, 38, 42–48 Advanced: 12–24, 29–48

REACHING ALL LEARNERS – Differentiated Instruction for students with

Developing Knowledge	On-level Knowledge	Advanced Knowledge	English Language Development
☐ Inclusion TE p. 757	☐ Cooperative Learning TE p. 757	☐ Cooperative Learning TE p. 757	☐ Cooperative Learning TE p. 757
☐ Practice A 14-5 CRB	☐ Practice B 14-5 CRB	☐ Practice C 14-5 CRB	☐ Practice A, B, or C 14-5 CRB
☐ Reteach 14-5 CRB	☐ Puzzles, Twisters & Teasers 14-5 CRB	☐ Challenge 14-5 CRB	☐ *Success for ELL* 14-5
☐ Homework Help Online Keyword: MT7 14-5	☐ Homework Help Online Keyword: MT7 14-5	☐ Homework Help Online Keyword: MT7 14-5	☐ Homework Help Online Keyword: MT7 14-5
☐ *Lesson Tutorial Video* 14-5	☐ *Lesson Tutorial Video* 14-5	☐ *Lesson Tutorial Video* 14-5	☐ *Lesson Tutorial Video* 14-5
☐ Reading Strategies 14-5 CRB	☐ Problem Solving 14-5 CRB	☐ Problem Solving 14-5 CRB	☐ Reading Strategies 14-5 CRB
☐ *Questioning Strategies* pp. 212–213	☐ Visual TE p. 75	☐ Visual TE p. 75	
☐ *IDEA Works!* 14-5			☐ *Multilingual Glossary*

ASSESSMENT
☐ Lesson Quiz, TE p. 759 and DT 14-5   ☐ State-Specific Test Prep Online Keyword: MT7 TestPrep

Teacher's Name _____ Class _____ Date _____

Lesson Plan 14-6
Multiplying Binomials pp. 762–765 Day _____

Objective Students multiply binomials.

> **NCTM Standards:** Develop and evaluate inferences and predictions that are based on data; Understand and apply basic concepts of probability; Recognize and apply mathematics in contexts outside of mathematics.

Pacing
☐ 45-minute Classes: 1 day   ☐ 90-minute Classes: 1/2 day   ☐ Other_____

WARM UP
☐ Warm Up TE p. 762 and Daily Transparency 14-6
☐ Problem of the Day TE p. 762 and Daily Transparency 14-6

TEACH
☐ Lesson Presentation CD-ROM 14-6
☐ Alternate Opener, Explorations Transparency 14-6, TE p. 762, and Exploration 14-6
☐ Reaching All Learners TE p. 763
☐ Teaching Transparency 14-6
☐ *Technology Lab Activities* 14-6
☐ *Know-It Notebook* 14-6

PRACTICE AND APPLY
☐ Example 1: Average: 1–6, 12–20, 29–31, 45–53 Advanced: 12–20, 33–35, 39, 45–53
☐ Example 2: Average: 1–7, 12–21, 29–31, 41, 45–53 Advanced: 12–21, 33–35, 39, 41–53
☐ Example 3: Average: 1–33, 36, 38, 40, 41, 45–53 Advanced: 12–53

REACHING ALL LEARNERS – Differentiated Instruction for students with

Developing Knowledge	On-level Knowledge	Advanced Knowledge	English Language Development
☐ Modeling TE p. 763	☐ Modeling TE p. 763	☐ Modeling TE p. 763	☐ Modeling TE p. 763
☐ Practice A 14-6 CRB	☐ Practice B 14-6 CRB	☐ Practice C 14-6 CRB	☐ Practice A, B, or C 14-6 CRB
☐ Reteach 14-6 CRB	☐ Puzzles, Twisters & Teasers 14-6 CRB	☐ Challenge 14-6 CRB	☐ *Success for ELL* 14-6
☐ Homework Help Online Keyword: MT7 14-6	☐ Homework Help Online Keyword: MT7 14-6	☐ Homework Help Online Keyword: MT7 14-6	☐ Homework Help Online Keyword: MT7 14-6
☐ *Lesson Tutorial Video* 14-6	☐ *Lesson Tutorial Video* 14-6	☐ *Lesson Tutorial Video* 14-6	☐ *Lesson Tutorial Video* 14-6
☐ Reading Strategies 14-6 CRB	☐ Problem Solving 14-6 CRB	☐ Problem Solving 14-6 CRB	☐ Reading Strategies 14-6 CRB
☐ *Questioning Strategies* pp. 214–215			☐ Lesson Vocabulary SE p. 762
☐ *IDEA Works!* 14-6			☐ *Multilingual Glossary*

ASSESSMENT
☐ Lesson Quiz, TE p. 765 and DT 14-6   ☐ State-Specific Test Prep Online Keyword: MT7 TestPrep